Lecture Notes
in Computational Science and Engineering

54

Editors

Timothy J. Barth
Michael Griebel
David E. Keyes
Risto M. Nieminen
Dirk Roose
Tamar Schlick

Jörn Behrens

Adaptive Atmospheric Modeling

Key Techniques in Grid Generation,
Data Structures, and Numerical Operations
with Applications

With 74 Figures and 3 Tables

 Springer

Jörn Behrens

Afred-Wegener-Institut
für Polar- und Meeresforschung
Am Handelshafen 12
27570 Bremerhaven, Germany
email: jbehrens@awi-bremerhaven.de

Library of Congress Control Number: 2006928282

Mathematics Subject Classification: 86A10, 76M10, 76M12, 76M20, 65Y05, 65M25, 65M50

ISBN-10 3-540-33382-7 Springer Berlin Heidelberg New York
ISBN-13 978-3-540-33382-8 Springer Berlin Heidelberg New York

Springer is a part of Springer Science+Business Media
springer.com
© Springer-Verlag Berlin Heidelberg 2006

Typesetting: by the author and techbooks using a Springer LATEX macro package
Cover design: *design & production* GmbH, Heidelberg

Printed on acid-free paper   SPIN: 11734871   46/techbooks   5 4 3 2 1 0

Für Katja, Laila und Janka.
Ihr seid das größte Glück in meinem Leben.

Preface

This work represents the essence of nearly 16 years of work in scientific computing for ocean and atmospheric modeling. This journey started at Alfred-Wegener-Institute for Polar and Marine Research in Bremerhaven, Germany, where – as a post graduate – I was assigned to optimize multi-grid solvers for elliptic partial differential equations evolving from ocean modeling. I am lucky and grateful that I had the chance to go all that way, to have the opportunity to explore the subject from many different angles, to have wonderful teachers and colleagues, and to have the chance to work and visit such exquisit places as the *National Center for Atmospheric Research* in Boulder, Colorado, USA, the *Frontier Research System for Global Change at the Yokohama Institute for Earth Science* (Earth Simulator) in Yokohama, Japan, the *Max-Planck-Institute for Meteorology* in Hamburg, Germany, the aforementioned *Alfred-Wegener-Institute for Polar and Marine Research* in Bremerhaven and Potsdam, Germany, the *Technische Universität München* in Garching, Germany, the *Fields Institute* in Toronto, Canada, the *Naval Research Laboratory* in Monterey, California, USA, the *Department of Informatics at the University of Bergen*, Norway and the *Centre for Mathematical Sciences at the University of Cambridge*, UK, to name only the most influential ones.

My own interest in adaptive methods arose from the exploration of finite element methods. When comparing finite elements on simple model problems with finite differences, the former method is often disregarded for computational efficiency and implementation complexity issues. However, finite element methods are much more versatile and flexible when it comes to irregular domains and locally refined meshes. Besides, finite elements are mathematically the more elegant approach. Being an aesthete, the wish for a method that fully unfolds the beauty of finite element methods was created. That was the start for the development of the adaptive semi-Lagrangian finite element method, better to be called adaptive Lagrange-Galerkin method [34]. Since then, my research was focused on adaptivity and the solution of geophysical fluid dynamics problems with advanced adaptive numerical methods. This was during my PhD period at Alfred-Wegener-Institute (AWI), where a prototype implementation of a shallow water solver was accomplished.

With a grant from the Federal Ministry of Education and Research (BMBF), I started the development of a parallelizable adaptive mesh generation tool for oceanic and atmospheric applications in the mid 1990's. At that time there was no such tool available. This work was continued after funding ran out by an internal grant from AWI, before I decided to enlarge my scientific background by changing to Technische Universität München (TUM). The time at TUM was great in that it provided me with a lot of new knowledge in numerical analysis. I owe my teacher Folkmar Bornemann a great debt of gratitude for his patience and his clearly structured and precise way of teaching.

By now, several groups have gained a lot of experience in adaptive modeling. Yet in atmospheric sciences, the number is still limited. To fulfil my requirements of a German Habilitation, I considered to just compose some of my articles and reports for a short written document of my work in adaptive atmospheric modeling. However, thinking again, I am now convinced that taking the chance of having to write a monograph, is the best excuse for doing this a bit more carefully and summarizing what has been done in adaptive atmospheric modeling so far. I am aware that this snapshot can only be incomplete. However, the references and approaches mentioned may at least give a good starting point for research in adaptive atmospheric modeling.

Acknowledgements

The work of the last few years has been supported by different sources, which I am grateful for:

- Bundesministerium für Bildung, Wissenschaft, Forschung und Technologie (BMBF), grant no. 07/VKV01/1.
- Alfred-Wegener-Institute (AWI) "Programm zur Förderung besonderer Forschungsthemen", title *Anwendungen der Multiskalenmodellierung mit adaptiven Finite-Elemente-Methoden.*
- BMBF, grant no. 01 LD 0037 within the DEKLIM research program.
- Deutsche Forschungsgemeinschaft (DFG), grant no. BE 2314/3-1.
- DFG travel grants, no's. BE 2314/1-1, BE 2314/2-1, and BE 2314/4-1.
- Deutscher Akademischer Austauschdienst (DAAD) PPP-Norway grant no. 02/29189 (with Tor Sørevik).

Without money no work can be done, but even more important are the people that had influence on my research.

Wolfgang Hiller of Alfred-Wegener-Institute, Bremerhaven, created the scientific freedom to start with the development of adaptive software for atmospheric modeling. The development of the software package `amatos` has been tremendously accelerated by *Natalja Rakowsky*, (Technische Universität

Hamburg-Harburg, formerly Alfred-Wegener-Institute, Bremerhaven), *Lars Mentrup, Florian Klaschka* (both Technische Universität München), *Matthias Läuter* (Alfred-Wegener-Institute Potsdam, he contributed figs. 8.9 and 8.21), *Stephan Frickenhaus* (Alfred-Wegener-Institute, Bremerhaven), and *Thomas Heinze* (Deutscher Wetterdienst, formerly TU München). I thank my students who contributed single pieces in the whole adaptive modeling puzzle.

I owe gratefulness to *Folkmar Bornemann* (TU München), who supported my work at TU München tremendously. *Tor Sørevik's* (University of Bergen, Norway) support and his suggestions to the first version of this manuscript are gratefully acknowledged. I obtained several reviews by *Joseph Egger, Frank Giraldo, Thomas Heinze, Matthias Läuter, Lars Mentrup, Volkmar Wirth,* and anonymous reviewers that improved the text tremendously.

Many fruitful scientific discussions as well as very nice non-scientific experiences are owed to my friend *Francis X. Giraldo* from Monterey, CA, USA. *Dave Williamson* made my visit at NCAR, Boulder, CO, USA, possible and his mentorship was very valuable to me. The inspiring atmosphere in *Annick Pouquet's* group, including *Aimé Fournier* and *Duane Rosenberg*, helped to clarify ideas and concepts. *Rich Loft, Stephen Thomas, Ram Nair, Amik St-Cyr* among others, all contributed to the scientifically fruitful stay at NCAR.

Finally, and most importantly, my family gave me the support and strength to keep the direction in all these years. I am grateful for the chances that I were offered and the experience I could gain.

Munich,   *Jörn Behrens*
April 2006

Contents

List of Figures

List of Tables

1

Introduction

Adaptive MODELING in atmospheric sciences has evolved to a state of maturity that it seems to be the right time to summarize what has been achieved so far and to sketch the near future of research directions. This work gives an overview of current approaches to adaptive atmospheric modeling. The author has included material and results cited from other sources in order to give a broader overview of the different approaches. It is clear that his own work is described in more detail, even if this might not in all cases be the mainstream in technological development. Many of the achievements of the author's group are reported in a way that tries to be as understandable as possible, yet detailed enough to be reproducible. However, if in doubt, readability was given the preference.

In this introductory chapter some motivation for aiming at adaptive atmospheric modeling is given. This is essentially a collection of arguments, the author came across defending his approach against traditionalists especially in the early 1990's when adaptive modeling was still a small academic exercise and widely disregarded. However, it is not only a chain of arguments, but also a reasoning, why it might be advisable to take the challenge of additional programming complexity, computational overhead and mathematical sophistication to achieve consistent and efficient adaptive models.

When collecting the references, publications and talks on adaptivity, it turns out that many people did in fact think about adaptive methods for atmospheric problems. So, we give a (certainly not concise) list of people and approaches in adaptive atmospheric modeling. As a historic remark on adaptive mesh refinement we refer to Babuška's pioneering work [15], which was not related to atmospheric modeling. He coined the term of *self-adaptation*, which terms a method that adapts itself to the solution properties, while it computes the solution. We will omit the prefix "self" and call such methods *adaptive methods*.

Fig. 1.1. Meteosat satellite image: infrared channel on March 24, 2005
(http://meteosat.e-technik.uni-ulm.de/meteosat/)

1.1 Why Adaptivity?

When looking at atmospheric phenomena, one almost always observes interaction of a variety of scales. Looking at a satellite image (fig. 1.1), one can observe fine scale structures like a cyclone over the Atlantic ocean with filamentary structures and fronts that comprise a relevant length scale of approximately 5 to 10 km, while the global scale that drives the dynamics is of five orders of magnitude larger. Another example is depicted in fig. 1.2. Here we see clouds forming behind orographic features (a mountain range). The mountain range covers an expansion of the order of 10 to 20 kilometers. The typical lee wave cloud has a scale of several kilometers while the high frequency perturbation visible in the cloud patterns at the top of the picture have length scales of only a few tens of meters.

Adaptivity can in fact help to resolve local (small) scales that interact with global scales in one consistent way. It is however paramount to understand, that an adaptive method is only as good as the refinement criterion that controls adaptivity. The criterion has to be able to capture and predict the occurrence of critical features, while these are evolving as sub-grid processes. On the other hand, an adaptive method cannot be better than the underlying model. If the model equations don't consider turbulence, even a fine grid will not be able to reproduce turbulence.

In contrast to locally refined fixed mesh methods, adaptive refinement can capture moving features, like meandering jets or moving fronts. Again, this requires an *a posteriori* dynamic refinement criterion. While prescribed refinement regions can rely on the knowledge of experts, adaptivity requires a formal description of the criterion for refinement.

Adaptive methods can help to speed up the computation by avoiding unnecessary calculations and saving memory. This is especially true for localized

Fig. 1.2. Lee waves visible in a cloud pattern at the morning sky over Arches National Park (Utah, USA)

phenomena like tracer clouds, point sources or isolated vortices. On the other hand, the adaptation control requires some overhead for the refinement criterion, grid management and irregular data access. To our experience, even if the saving is not worth the effort time-wise, in most cases memory requirements are drastically reduced.

Once a balanced adaptive method is established, it is probably also very efficient on static or even uniform grids. So, in cases where no localized phenomena are expected, adaptivity control can be switched off, reducing the mentioned overhead. Still, these tools are available and can be used in the course of model simulation or analysis. With these methods at hand, one has the possibility to make a quantitative assessment of the quality of solution. Moreover, one is able to control certain inevitable inaccuracies. For example, by increasing the order of approximation (refinement), one can reduce the numerical dissipation of a method (locally) in order to avoid unwanted effects that spoil the solution. An adaptive method also helps in cases, where normally statically refined grids would be used. The user could use the adaptive control during a start-up phase and switch off adaptive control as soon as the method has converged to a stable grid spacing. By this, it can be guaranteed that the region of refinement is optimally chosen.

Irregular boundary shapes often influence the solution of equations describing atmospheric motion. On the other hand, irregular boundary features like mountains, valleys, shore lines, etc. introduce interesting atmospheric patterns that have to be resolved, when accurate modeling is aimed at. Many non-adaptive modeling techniques are not capable of resolving these features well enough. With an adaptive mesh refinement strategy, it is no additional effort to define an orography based refinement criterion that can capture small scale topographic features. An example of this is shown in fig. 1.3, where the shore of the shelf ice can be clearly identified by the refined grid.

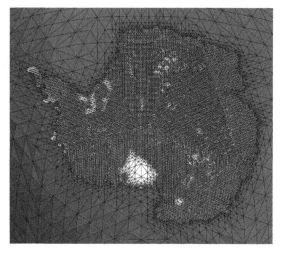

Fig. 1.3. Locally and statically refined grid over Antarctica, the topography gradient has been used as a refinement criterion

It is the author's vision that we will be able to develop numerical methods that are robust and efficient enough to be combined with adaptivity in standard (operational) environments in atmospheric research. When we look at the development of numerical methods for ordinary differential equations (see for example [120, 121, 182, 183]) we see that today it is just a standard to use adaptive integrators. These methods have reached a state of maturity that nobody really thinks about grids and adaptive control anymore, but the user decides on accuracy and the software fulfills the needs. This is the vision of adaptive atmospheric modeling (and probably adaptive solution of PDEs in general).

Thus, adaptivity triggers the concern about accuracy and local error much more than non-adaptive methods do. Once there is control over local approximation order, one has to decide on a tolerable error measure. At that point the question arises on what "error" really means in a fully equipped atmospheric model.

To summarize, the promise of adaptive atmospheric modeling is the

- locally accurate
- efficient
- error controlled
- truly multi-scale

simulation of atmospheric processes.

1.2 Who's Who in Adaptive Atmospheric Modeling

Early approaches to adaptive mesh refinement in atmospheric modeling were presented by researchers in Hurricane or Typhoon prediction that nested finer meshes into coarse mesh models of the large scale simulation. One example of this not yet adaptive approach is documented in [266]. Others are given in [247, 248]. Nesting is still a wide spread technique for achieving local high resolution [157]. However, this review is concerned with adaptive methods, which means methods that dynamically adapt to flow features during run time.

Early truly adaptive atmospheric models were developed by Klemp and Skamarock and by Dietachmayer and Droegemeier [125, 367, 369]. Klemp and Skamarock based their refinement strategy on a truncation error estimate [275].

There are two major adaptation principles, discussed in sect. 2.4. One does not change the grid topology (i.e. the number of grid points and the inter-connectivity) but changes the spacing between grid points by transformation functions [70]. The other principle refines/coarsens the mesh by inserting/deleting grid points and re-meshing (locally). Examples for the first approach are given in [125] and more recently in [77, 215]. The author's work and much of what follows in this book, follows the second approach.

The first – and to the author's knowledge only – adaptive and operational weather and dispersion model OMEGA uses a finite volume approach on an unstructured Delaunay mesh that is locally modified and smoothed [20, 174]. OMEGA has been developed by Bacon and coauthors at Science Applications International Corp., McLean, VA. Giraldo based at Naval Research Lab in Monterey, CA, has used Delaunay meshes for adaptively refined Lagrange-Galerkin methods to solve the shallow water equations [166]. Recently, a nodal spectral element method for triangular adaptively refined meshes has been proposed [171]. Iselin and coauthors published a dynamically adaptive version of the MPDATA scheme [215, 214]. MPDATA is a finite difference scheme that has several advantageous characteristics like conservation properties. Prusa and Smolarkiewicz combine a dynamic grid adaptation (movement) method with an Eulerian or semi-Lagrangian conservative advection scheme in [330]. Recently, Smolarkiewicz and Szmelter have presented an unstructured grid formulation of MPDATA [372]. There are several groups involved in adaptive air quality modeling [373, 240, 314].

Jablonowsky's dissertation, defended at the University of Michigan, is an example of a finite volume scheme on quadrilateral adaptively refined meshes applied to the 2D shallow water equations and the 3D baroclinic equations on the sphere [221]. Additionally, she gives a good introduction to adaptive methods in atmospheric modeling. Barros and Garcia recently introduced a variable resolution semi-Lagrangian model for global circulation with the shallow water equations [26]. Fiedler uses dynamic grid adaptation to resolve the boundary layer [146]. A wavelet approach has been utilized in adaptive ocean modeling in [223]. And very recently Rosenberg and coauthors demonstrated a parallel adaptive method to study geophysical turbulence problems [347].

In the UK a strong adaptive tradition in atmospheric modeling is represented by publications of Tomlin and coworkers. A fully 3D approach on unstructured tetrahedral meshes is documented in [394] and [163]. Hubbart and Nikiforakis use a 3D quadrilateral locally refined mesh for simulating global tracer transport in [207]. For the ocean Pain, Ford, Piggott and coworkers developed an advanced adaptive modeling tool for 3D unstructured computations [151, 152, 308, 321].

Ivanenko and Muratova introduce a quadrilateral distorted grid based finite difference shallow water model with local mesh refinement [220]. Blayo and Debreu use local mesh refinement in the context of finite difference methods for ocean modeling [61]. Kashiyama and Okada use adaptive mesh generation techniques for solving shallow water flow problems [236]. Again in ocean modeling an adaptive 3D shallow water model has been proposed by Ham and coworkers in the Netherlands [185]. Hundsdorfer et al. proposed an adaptive method of lines based finite difference approach for atmospheric transport problems [212].

In Germany the author's group is active since the early 1990's in adaptive atmospheric modeling [33, 34, 35, 36, 39]. Läuter introduced a shallow water model in vorticity-divergence formulation on the sphere [255, 256]. Hess has developed an adaptive and parallel multi-grid solver for the shallow water equations [201].

To summarize, research in adaptive atmospheric (and ocean) modeling is gaining momentum in recent years. This is documented in a study on numerical schemes for non-hydrostatic models that explicitly addresses suitability of schemes to adaptively refined meshes [376]. However, the group of people working on adaptive modeling in this field is still moderate, even if one takes into account that the above list is certainly not complete.

1.3 Structure of the Text

This work is intended to give an overview of adaptive atmospheric modeling techniques in a way that makes it possible to extract the basic ideas for the reader's own development, together with additional pointers to the more detailed literature. Several aspects are covered in more detail. At the end of this book, a reader should be able to draft a simple adaptive atmospheric modeling technique for his/her own needs and find literature with detailed descriptions on the construction.

Chapter 2 introduces several principles of adaptive modeling. It is of a more abstract and basic character, since adaptive methods are not well established in the atmospheric modeling community yet.

In chap. 3 grid generation techniques are covered in some detail. Grid refinement is the major tool for multi-scale adaptivity in atmospheric modeling to date. Therefore, some effort has been put into the compilation of 2D and 3D triangular (unstructured) and quadrilateral (structured) mesh refinement strategies. Additionally basic initial grids for spherical geometries are introduced.

Computational aspects are treated in chap. 4. Grid generation – especially for unstructured grids – needs adequate data structures. On the other hand, numerical algorithms need consecutive data structures for efficient execution. Both requirements are covered there.

Chapter 5 treats issues in parallelization. Efficient execution on modern computing architectures calls for parallelizable data structures and dynamic load balancing. In this regard, chaps. 4 and 5 are interrelated. While the description in the former is valid for uniprocessor architectures alike, the latter is only valid for parallel architectures.

Chapters 6 and 7 deal with the more mathematical aspects of adaptive modeling: the representation of differential operators on adaptive grids and the discretization of conservation laws. While conservation laws are the basic equations relevant in atmospheric modeling, they are mostly given in terms of partial differential equations. Thus, the discretization of conservation laws requires differential operators to be represented on adaptive meshes.

Chapter 8 gives examples of the successful application of adaptive methods to atmospheric modeling. Additionally, test cases for studying basic properties of adaptive methods are introduced.

In Chap. 9 we draw conclusions from the material in the previous chapters and try to derive basic rules of good practice. Finally, we try to pave the way to the future of adaptive techniques in atmospheric modeling. It is the authors belief that adaptivity is a basic principle of multi-scale processes. Therefore, adaptive techniques will help to understand and solve future problems in atmospheric modeling.

In the appendix, we collect some of the basic mathematical tools and more technical descriptions that hindered the fluent reading in the chapters described above.

2

Principles of Adaptive Atmospheric Modeling

In this chapter, major principles of adaptivity and especially adaptive atmospheric modeling will be discussed. We start with reconsidering the paradigms of adaptivity in the mathematical and in the meteorological communities respectively in order to clarify the basic notations. After this clarification, and a description of principle challenges in adaptivity, we introduce concepts of adaptive refinement techniques and discuss refinement criteria.

2.1 Paradigms of Grid Refinement – Resolution Enhancement Versus Error Equilibration

A common cause for misunderstanding when talking about adaptive mesh refinement, or adaptivity in general, is the existence of two different approaches. In order to avoid ambiguity or confusion this section explains the two paradigms.

The first approach is interested in increasing resolution locally. This is the prevailing aim in most meteorological applications, and fixed local mesh refinement has long been used to achieve this goal (see for example [101, 117]). The approach is driven by the physical understanding that increased resolution reveals more structure in the simulated phenomena.

An example for this approach is depicted in fig. 2.1. The locally refined mesh region follows the tracer constituent in a trace gas advection and dispersion simulation. By this measure, the number of unknowns can be kept low, while maintaining high resolution, where this is necessary to capture fine scale features (in this case the point source and the sharp front of the tracer cloud). In fact, the adaptive computation needs between an order of 10^3 and 10^4 cells in contrast to an order of $2 \cdot 10^5$ cells for a uniformly refined mesh of same resolution.

The other approach is motivated from the mathematical side. Assuming that a well defined partial differential equation is given, adaptive refinement

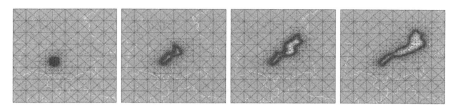

Fig. 2.1. Locally adapted meshes for tracer dispersion experiment over Southern GERMANY; FOUR time steps are shown, starting from a point source

of the approximation order or mesh width intends to equilibrate the numerical error. This approach requires that a consistent formulation of the error is available and also that a capable estimation of that error is feasible. It also requires that the numerical method together with the discretization of the problem are both convergent. This might not be the case in real life meteorological applications, where discrete models of sub-grid processes can destroy convergence.

This paradigm was introduced in the early 1970's probably by Babuška [15, 18]. It was originally used with finite element methods (FEM). For a historical draft of the development of FEM see [90, 390]. An example is depicted in fig. 2.2, namely the solution of Poisson's equation on a circular domain with a Dirac delta function load (right hand side). This example is taken from the MATLAB PDE toolbox demos [277].

In many cases the two approaches will find similar solutions. Since intuitively, areas of high physical activity will be those of high mathematical error (e.g. steep gradient regions, high local curvature of some constituent). However, there are cases, where the mathematically induced error-driven criterion refines counter-intuitively. This statement is especially valid for transport dominated problems, found in atmospheric modeling, since error can be propagated and the propagation direction may not be predictable. In general, refinement criteria are very problem dependent and have to be chosen after a careful analysis of the problem's requirements (see sect. 2.6 for a more detailed description of refinement criteria).

2.2 Principle Difficulties with Adaptivity

While an adaptive method promises to be more efficient for certain cases of localized features, it incurs several new difficulties that need to be treated with care. Some difficulties can be circumvented by reasonable choices of software methods, others need to be tackled by mathematically more sophisticated methods. This section lists some of the principle difficulties that will be treated in the course of this presentation.

Let us first look at the computer science part of adaptive methods. Since locally refined meshes are mostly unstructured, indirect addressing and the

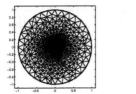 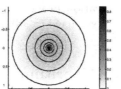 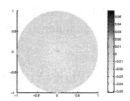

Fig. 2.2. Locally adapted mesh for error equilibration: (**left**) grid, (**center**) solution, (**right**) error $|u - u_h|$; figures created by a MATLAB^{TM} PDE toolbox demo

corresponding performance loss is an issue. Even in the case of locally structured grids, an overhead of address information has to be taken into account.

When dealing with adaptively refined meshes, advanced software technology has to be used for the dynamically changing data structures. Most adaptive programs utilize object oriented data structures for grid handling. While object orientation is efficient for grid manipulations, it is not well suited for high performance numerical calculations. To achieve high performance, consecutive data structures like arrays or (blocked) matrices are needed.

When applying adaptivity to time-dependent problems, re-meshing may occur in almost every time-step. This results in load-imbalance for parallel computations. So, efficient dynamical load balancing schemes are necessary in order to maintain parallel efficiency of adaptive schemes.

There are also principle difficulties concerning the mathematics of adaptive methods. For explicit time-stepping schemes, the grid size of the finest mesh level dominates the CFL stability criterion. Therefore, either locally adaptive time sub-stepping or an unconditionally stable time-integrator is required, in order to avoid instability.

The most demanding difficulty for adaptive methods is to find a suitable refinement criterion. In general, an adaptive mesh refinement algorithm is only as good as the criterion for refinement. Thus, if one aims at substantial improvement in accuracy of the solution, a good understanding is needed for what accuracy really means in context of the problem and full insight into the reason for error is required. An effective refinement criterion needs to detect those areas that cause the highest error or that are of highest interest, concerning the relevant physical features.

When dealing with different orders of convergence in the same application, sub-grid scale process parameterizations need to be tailored such that they are independent of grid size or polynomial order. Up to now, the interleaving of continuous modeling and discretization has lead to parametrization that is optimal to a fixed mesh.

When coupling an adaptive method to other external models or post processing systems, data often have to be represented on structured regular grids. Interpolation onto such interface grids needs to be consistent with the modeling error and conservation properties of the physical constituents that are communicated (e.g. fluxes have to adhere to mass conservation).

2.3 Abstract Adaptive Algorithm

In this section we will introduce an abstract formulation of the adaptive algorithm for atmospheric transport-dominated problems. This algorithm is based on *a posteriori* refinement criteria. We will also define basic notations.

It is important to distinguish between *a priori* and *a posteriori* error control. While a priori refinement criteria where used even in the very first paper describing finite element methods [397], a posteriori error control is a later concept.

With *a priori error control* or *a priori refinement* criteria we denote those methods that have prior knowledge about the solution of the problem and react by refining the approximation accordingly. Examples of this are models that refine over areas of interest [101, 117, 287], or in areas of known high turbulent activity, e.g. near steep topographical features (mountains, sea mounts, valleys, etc.) [91, 107, 152, 185, 426].

For an adaptive algorithm, some kind of a posteriori refinement criterion is necessary. The algorithm starts with solving the problem on a provisional grid. Then, a posteriory (i.e. after solving), a criterion determines those regions of the computational domain that need to be refined. Finally, a new solution is computed on the refined grid.

In general the true error is not available during the computation. So, for a mathematical determination of refinement areas, i.e. areas of large local error, an error estimator is needed. The concept of *self-adaptation* has been coined by Babuška [15]. The mathematical error estimation approach requires deep theoretical insight into the nature of error of the numerical method together with the problem to be solved.

A posteriori refinement control can also be derived from heuristical criteria. In many cases gradients of relevant physical quantities or curvature properties are used for refinement. It is also possible to use proxy data in order to obtain a refinement criterion. So, even without deep theoretical error analysis, and even without a clear concept of error, an abstract a posteriori adaptively refined algorithm can be formulated.

Before stating the adaptive algorithm, the basic non-adaptive algorithm should be formulated. For transport dominated atmospheric flow problems, it is assumed that a discretization of the governing equations is performed in space and time separately. Thus, we first semi-discretize in time to obtain a time-stepping scheme, and then discretize the remaining stationary problem in space.

Let us assume that the time interval of interest I is normalized to start at time 0, $I = [0, T] \in \mathbb{R}$. Let us denote the spatial domain with $\mathcal{G} \subset \mathbb{R}^d$ and denote ρ_h the discrete variable corresponding to the continuous variable ρ. In general, ρ defines a map $\rho : \mathcal{G} \times I \to \mathbb{R}^r$, where $r = 1$ for scalar and $r = d$ for vector-valued variables. Then ρ_h defines a map $\rho_h : \mathcal{G}_h \times I_h \to \mathbb{R}^r$. Here $\mathcal{G}_h$ is some discrete representation of $\mathcal{G}$, for example a polygonal approximation of the true domain, which can be discretized by a triangulation $\mathcal{T} = \{\tau_1, \ldots, \tau_M\}$ (see sect. 3.1 for a formal definition of a triangulation). I_h is the discretized time interval $I_h = \{t_i : 0 \le t_i \le T(i = 0 : K)\}$. If we deal with systems then ρ and ρ_h represent vectors of all prognostic variables. We formulate the abstract problem by the following equation

$$\frac{\partial}{\partial t}\rho + \mathcal{D}\rho = r, \tag{2.1}$$

with a differential operator $\mathcal{D} = \mathcal{D}(t, \mathbf{x}, \rho)$ and a right hand side $r = r(t, \mathbf{x}, \rho)$. Additionally initial values $\rho(t = 0) = \rho_0$ and boundary values $\rho(t)|_{\partial\mathcal{G}} = \rho_b(t)$ are assumed to be given. Furthermore, we assume that the time derivative can be represented as an evolution

$$\rho(t > 0) = \Psi(0, t; \rho(0)),$$

and that the stationary problem can be solved separately and is represented by

$$\mathcal{D}\xi = r,$$

where ξ can then be used in the evolution, $\rho(t > 0) = \Psi(0, t; \xi)$. With these assumptions, the basic non-adaptive algorithm reads as follows

Algorithm 2.3.1 *(Basic non-adaptive algorithm)*

1. *initialize prognostic variables ρ_h, i.e. set values at all relevant time levels $t_i < t_0$: $\rho_h(t_i < 0) = \rho_0(t_i)$;*
2. **FOR** *all time steps $t_i \in I_h$, $i = 1 : K$* **DO**:
3. *integrate problem in time interval $[t_{i-1}, t_i]$:*
   a) *evaluate discrete differential operators to obtain right hand side*

$$r = r(t \le t_{i-1}, \mathbf{x}(t \le t_{i-1}), \rho_h(t \le t_{i-1})),$$

   depending only on variable values at times $t \le t_{i-1}$;
   b) *Solve stationary problem*

$$\mathcal{D}_h\xi_h(t_i) = r$$

   with ξ_h an intermediate representation of the constituents;
   c) *update prognostic variables at time t_i:*

$$\rho_h(t_i) = \Psi_h(t_i, t_{i-1}; \xi_h(t \le t_{i-1})),$$

   where Ψ_h is the discrete evolution operator;
4. *perform any other diagnostic calculations depending on values $\rho_h(t_i)$;*
5. **END DO**.

Remark 2.3.2 *Note that algorithm 2.3.1 tries to be rather general. For example, an explicit method would have a diagonal system matrix $S_h = \text{diag}(\mathbf{s}_h)$, where $\mathbf{s}_h$ is a vector of coefficients. However, this algorithm does not cover all possible discretization schemes for PDEs as they occur in atmospheric modeling, because it assumes that the discretization leads to a time stepping scheme and that an evolution operator can be derived from the system of equations.*

With this formal abstract definition of a time-stepping algorithm, it requires just two additional steps to define a basic adaptive algorithm:

Algorithm 2.3.3 *(Basic adaptive algorithm)*

1. *initialize prognostic variables ρ_h (as in algorithm 2.3.1);*
2. **FOR** *all time steps $t_i \in I_h$, $i = 1 : K$* **DO:**
3. *integrate problem in time interval $[t_{i-1}, t_i]$:*
   a) evaluate discrete differential operators to obtain right hand side

$$r = r(t \le t_{i-1}, \mathbf{x}(t \le t_{i-1}), \rho_h(t \le t_{i-1})),$$

   depending only on variable values at times $t \le t_{i-1}$;
   b) Solve stationary problem

$$\mathcal{D}_h \xi_h(t_i) = r$$

   with ξ_h an intermediate representation of the constituents;
   c) update prognostic variables at time t_i:

$$\rho_h(t_i) = \Psi_h(t_i, t_{i-1}; \xi_h(t \le t_{i-1})),$$

   where Ψ_h is the discrete evolution operator;
   d) calculate local refinement criterion η_τ for each $\tau \in \mathcal{T}$, and refine those τ, where $\eta_\tau > \theta_{\text{ref}}$, with θ_{ref} a given refinement tolerance;
   *e) **IF** grid changed, **THEN:** return to step 3a, **ELSE:** continue with step 4;*
4. *perform any other diagnostic calculations depending on values $\rho_h(t_i)$;*
5. **END DO.**

Note that only steps 3d and 3e in algorithm 2.3.3 have been added to the original basic algorithm 2.3.1. However, in these two steps, a huge amount of computational intelligence is involved. We will specify the details of the refinement strategy in sects. 2.4 and 2.5 below. Furthermore, grid refinement in geometric terms is discussed in chap. 3, while some remarks on refinement criteria are made in sect. 2.6.

2.4 Dynamic Grid Adaptation – *h*-Refinement

One principle of adaptivity is to refine the computational mesh. This will be the predominant method throughout this text. Commonly, the spatial mesh size is denoted by h, therefore, refining the mesh corresponds to refining h. So, a common name for *adaptive mesh refinement* (AMR) in the mathematical community is *h-refinement*.

h-refinement is based on an approximation property of the kind

$$\varepsilon_h = \|\rho - \rho_h\| \leq Ch^\nu \|\rho\|, \tag{2.2}$$

where we denote by ρ the true solution of some given problem, ρ_h is a discrete approximation, C is a constant, h is a parameter indicating the mesh width, and ν is the order of convergence. Equation (2.2) states that for $h \to 0$ the discrete solution ρ_h tends to the true solution ρ or in other words, the error $\varepsilon_h \to 0$.

In the following we want to denote by

- ε_h the discretization error: $\varepsilon_h = \|\rho - \rho_h\|$ in an appropriate norm;
- ε_τ the local discretization error $\varepsilon_h|_\tau$;
- $[\varepsilon_h]$, and $[\varepsilon_\tau]$ an estimation of the global and local discretization error, resp.;
- η_τ or η_i for each τ_i, a local criterion that can be an error estimate or some other local measure that can be used to derive a refinement.

There are several strategies to select the refinement regions. Ultimately, one is interested in getting the most accurate solution with least effort. Therefore, an algorithm for refinement has to be given. For now, let us assume that we can compute a local error, denoted by ε_τ (we will present common error estimators and refinement criteria in sect. 2.6). A first simple approach is based on the maximum norm of the local error vector $\varepsilon_m := (\varepsilon_{\tau_1}, \ldots, \varepsilon_{\tau_m})$. With $\varepsilon_h = \varepsilon_{\max} = \|\varepsilon_m\|_\infty = \max_{k=1:m} \varepsilon_{\tau_k}$ the global error in the ∞-norm, one can try to minimize the global error by refining those elements of the mesh that have largest local errors.

More precisely, with a given tolerance $0 \leq \theta_{\text{ref}} \leq 1$, a simple algorithm for local mesh refinement is given by

Algorithm 2.4.1 *(∞-Norm based h-refinement)*

1. *Let $\mathcal{T}_0$ be a given initial triangulation, ε^0 the vector of local error components ε_{τ_i}, $i = 1 : m$, and $\varepsilon_{\max}^0 = \|\varepsilon^0\|_\infty$.*
2. FOR *all $\tau \in \mathcal{T}_0$* DO:
3. IF *$\varepsilon_\tau > \theta_{\text{ref}}\varepsilon_{\max}^0$* THEN:
   a) *refine τ*
4. END DO.

While algorithm 2.4.1 is scaling invariant (i.e. if ε^0 is multiplied by a scalar α, then the result of refinement does not change) and extremely simple, there is in fact no theoretical guarantee, that the discretization error is always reduced, or that an algorithm equipped with this refinement procedure really converges. However, in practical applications this strategy has been used successfully (see [39] for its application in a tracer transport problem).

Another p-norm based approach is the so called *equi-distribution strategy* [139]. It is best illustrated by assuming a uniform distribution of local errors $\varepsilon_\tau = \bar{\varepsilon}$ for all $\tau \in \mathcal{T}_0$. Now, we define the global error to be the p-norm of the local error vector:

$$\varepsilon = \|\varepsilon_m\|_p = \left(\sum_{k=1:m} \varepsilon_{\tau_k}^p \right)^{\frac{1}{p}} = (m\bar{\varepsilon}^p)^{\frac{1}{p}} = m^{\frac{1}{p}}\bar{\varepsilon}.$$

We require $\varepsilon \le \theta_{\text{ref}}$. If we let $\varepsilon = \theta_{\text{ref}}$, we can see that $\bar{\varepsilon} = \frac{\theta_{\text{ref}}}{m^{\frac{1}{p}}}$. From this, we derive the heuristic

Algorithm 2.4.2 *(equi-distribution h-refinement)*

1. *Let $\mathcal{T}_0$ be a given initial triangulation.*
2. FOR *all* $\tau \in \mathcal{T}_0$ DO:
3. IF $\varepsilon_\tau > \frac{\theta_{\text{ref}}}{m^{\frac{1}{p}}}$ THEN:
   a) *refine* τ
4. END DO.

Läuter uses this strategy for $p = 2$ in an application to adaptive shallow water modeling in [256].

Dörfler introduces a guaranteed error reduction strategy [128]. This strategy – only outlined here, since it is hardly ever used in atmospheric modeling so far – is based on the idea that a subset of mesh elements is refined with the sum of their local errors being a fixed part of the total error ε. Thus, given a parameter $0 < \theta < 1$, find a minimal set $S = \{\tau_1, \ldots, \tau_k\} \subseteq \mathcal{T}_0$, such that

$$\sum_{\tau \in S} \varepsilon_\tau^p \ge (1 - \theta)^p \varepsilon^p. \tag{2.3}$$

It follows that if S is being refined, the error will be reduced by a factor depending on θ and the properties of the problem data (right hand side, etc.). Dörfler proposes a strategy for selecting the elements in S by an inner iteration. In this inner iteration, the ∞-norm strategy 2.4.1 is used with a decreasing threshold θ_{ref} until S is large enough to meet the requirement in (2.3).

So far, only refinement strategies have been considered. In transport dominated atmospheric modeling the area of interest may move over time, so that we may need to coarsen the mesh after a while. Both of the above strategies 2.4.1 and 2.4.2 can be inverted easily to be used as coarsening strategies. As an example, we demonstrate this for the ∞-norm strategy 2.4.1. Let $0 \le \theta_{\text{crs}} < \theta_{\text{ref}}$. Then one may coarsen an element τ of the triangulation if

$$\varepsilon_\tau < \theta_{\text{crs}} \varepsilon_{\text{max}}^0.$$

There are two topics that have to be kept in mind. First, the choice of θ_{crs} has to be done carefully in order to avoid incompatible refinement and coarsening conditions. These in turn can cause oscillating refinement/coarsening of single elements. In mathematical a posteriori error estimation, one can usually give an approximation of ε_τ in terms of the element (mesh) width h_τ:

$$\varepsilon_\tau \leq Ch_\tau^\nu,$$

where ν is an appropriate exponent. For linear approximation order ($\nu = 1$), coarsening the element such that the mesh width is doubled, causes the local error to increase by a factor of 2. Therefore, we have to choose

$$\theta_{\mathrm{crs}} \leq \frac{\theta_{\mathrm{ref}}}{2},$$

in order to avoid oscillatory behavior. For higher approximation order this argument has to be modified appropriately.

Secondly, we need to define a coarsening strategy, since an element can only be coarsened, if the corresponding siblings are also coarsened. This strategy is described in sect. 3 in connection with the refinement strategies.

2.5 Adapting the Order of Local Basis Functions – p-Refinement

In contrast to the h-refinement of the previous section, where the grid is refined, one can also increase the order of polynomial approximation, while leaving the mesh unchanged. This p-refinement, where p stands for the order of approximation, has been developed in the context of finite element methods in the early 1970's [19, 382]. The foundation for the p-refinement is given by a similar approximation property as (2.2). Let $\mathcal{T}_0$ be a given (and fixed) triangulation of the computational domain $\mathcal{G} \subset \mathbb{R}^d$, and let p be the order of approximation. Then at typical error approximation holds

$$\varepsilon_p = \|\rho - \rho_p\| \leq Cp^\nu \|\rho\|, \tag{2.4}$$

where ρ_p denotes an approximation of order p to the true solution ρ and – as in (2.2) – ν is an appropriate exponent related to the convergence and C is a constant.

In the context of finite elements let $\mathcal{P}_p(\mathcal{T}_0)$ be the space of continuous functions that are piecewise polynomials of degree p in each element $\tau \in \mathcal{T}_0$ and globally continuous. Then for $\rho \in H^k(\mathcal{G})$ there exists a sequence $\rho_p \in \mathcal{P}_p(\mathcal{T}_0)$, $p = 1, 2, \ldots$ such that for $0 \leq l \leq k$

$$\|\rho - \rho_p\|_l \leq Cp^{-(k-l)} \|\rho\|_k. \tag{2.5}$$

For a proof of (2.5) see [19]. Equation (2.4) – or similarly (2.5) – states that by increasing the order of polynomial approximation the error is minimized, $\varepsilon_p \to 0$.

In order to create a sequence of approximations of increasing order, a hierarchy of supporting functions needs to be constructed. In finite element analysis, these hierarchical basis function families are easily constructed starting from the well known linear basis function [89]. For demonstration purposes, we give an example of the first three orders of basis functions for 2D triangular C^0 finite elements here [382]. For the linear (order $p = 1$) element, we define the basis functions b_i, $i = 1 : 3$ by

$$\begin{aligned}
b_1(x_1, x_2) &= 1 - (x_1 + x_2), \\
b_2(x_1, x_2) &= x_1, \\
b_3(x_1, x_2) &= x_2,
\end{aligned} \tag{2.6}$$

where we assume b_i to be defined on the unit triangle τ_1 being the convex hull of the three vertices $\{(0,0), (1,0), (0,1)\}$.

Now, the second order basis functions are composed hierarchically, by the first order basis functions and three additional basis functions normalized by a factor of $-\frac{1}{2}$ and evaluating the second derivatives of the approximating polynomials at the vertices (since the approximating polynomials are quadratic, the second derivatives are constant). The basis functions are given by

$$\begin{aligned}
b_4 &= b_1 b_2, \\
b_5 &= b_2 b_3, \\
b_6 &= b_3 b_1.
\end{aligned}$$

Finally, the third order basis functions can be constructed by re-using the first and second order basis functions and adding four more terms: the evaluations of the three third derivatives at edge midpoints, a normalizing factor of $\frac{1}{12}$, and an internal node value at the barycenter of the triangle. With this additional information the basis functions are given by

$$\begin{aligned}
b_7 &= b_1^2 b_2 - b_1 b_2^2, \\
b_8 &= b_2^2 b_3 - b_2 b_3^2, \\
b_9 &= b_3^2 b_1 - b_3 b_1^2, \\
b_{10} &= b_1 b_2 b_3.
\end{aligned}$$

In the early papers, authors showed superior properties of the p-refinement over the h-refinement. However, increased computing power and principle difficulties to increase the order of approximation arbitrarily, helped the h-refinement or AMR methods to become dominant. p-refinement is still a viable and very promising option in the hp-refinement methods, where one attempts to balance either mesh refinement or increase of order by appropriate error analysis techniques [16, 303, 321, 358]. A combination of p-refinement in a moving mesh method and additional h-refinement is given in [251]. A comparison of h and p refinement for solutions of the shallow water equations has been conducted in [405].

2.6 Refinement Criteria

Refinement criteria are crucial to the whole process of adaptive modeling. The adaptive model can only be as good as the adaptation criterion. If the criterion over-estimates the error, in other words if it is too selective, then it detects too large refinement areas, and the method cannot be efficient. On the other hand, if the criterion under-estimates the error, the method will be inaccurate. So, the quest is to find a criterion that detects the refinement area as accurately as possible in order to maximize efficiency and accuracy of the adaptive method.

There are three basic types of refinement criteria:

1. Heuristic error proxies;
2. Physics based criteria;
3. Mathematical (discretization) error estimators.

Mathematical error estimators are often based on the residual of a given operator. These estimators developed along with the mathematical adaptivity paradigm of equilibrating the discretization error (see sect. 2.1). Physics based criteria are more common on the application side of adaptive modeling. They represent the perspective of resolution enhancement by adaptive refinement. Error proxies often bridge the gap between these two, since they represent a heuristical way to estimate the error by means of easily accessible (physical or functional) values. This section shall present several examples of the above types.

2.6.1 Error Proxies

Error proxies are criteria that are derived from easily available data. For example, it is known that steep orography gradients create complex flow patterns. Since the orography is available *a priori*, these gradients can be easily computed and be used to refine locally.

In tracer advection the gradient of the advected constituent can be used as a refinement criterion. One can argue that especially for semi-Lagrangian methods, the accuracy of interpolation drops where the gradient is steep. Therefore a gradient-based criterion can be used successfully for semi-Lagrangian advection (see sect. 8.1.2, and [34, 39]). For purely advection modeling, Kessler showed that a gradient based criterion results in even better approximation quality than a truncation error estimate [240, 393].

In general, proxy criteria often interleave with physics based criteria. A gradient or a curvature of a constituent is only a proxy, since not the gradient of the constituent causes the real problem but the low approximation quality of a numerical scheme. Karni and coworkers use a smoothness indicator for adaptive refinement control in a solver for hyperbolic systems [230].

In multi-constituent chemical transport modeling, a proxy based refinement, where the criterion is composed of several individual proxies, is indispensable. Tomlin et al. use such a weighted sum of proxy refinement indicators in [394]. Belwal et al. demonstrate a similar principle in [47].

2.6.2 Physics Based Criteria

Many physics based criteria have been developed in the context of cyclone tracking in nested or adapting hurricane modeling. For example the 850 hPa relative vorticity extreme values have been used to track cyclones. Successful approaches applied geopotential or temperature extremes as well as sea level pressure (see [221] for a review of methods). Läuter uses the direct values of vorticity and divergence to derive an efficient and accurate refinement criterion. More precisely, the refinement criterion η_τ in each cell τ of the triangulation is computed from vorticity ζ and divergence δ by

$$\eta_\tau = \left(\int_\tau \zeta^2 + \delta^2 \, dx \right)^2 = \|\zeta + \delta\|_{L^2(\tau)}.$$

The grid is refined, wherever η_τ is above a user defined threshold, that is scaled with an absolute error factor derived from the previous time step [256]. Different strategies for marking elements for refinement have been introduced sect. 2.4.

Jablonowski assesses three different physical adaptation criteria, the (relative) vorticity ζ_r, the gradient of the geopotential height field $\nabla\Phi$ and the curvature of the geopotential height, in other words the Laplacian $\Delta\Phi$ [221]. Her conclusions result in an error criterion based on either $\nabla\Phi$ or ζ_r, as in Läuter. However, she reports on alternating refinement/de-refinement behavior in her simulations as well as mislead refinement by high frequency oscillations in the solution due to numerical (instability) effects.

2.6.3 Mathematical Error Estimation – Basic Principles

Most mathematical error estimators rely on the residual of the differential operator with a discrete solution. We want to briefly outline the principle on an abstract example equation:

$$\mathcal{D}(\rho) = f \text{ in } \mathcal{G} \subset \mathbb{R}^d. \tag{2.7}$$

$\mathcal{D}$ represents a given linear (differential) operator and f a right hand side; we assume suitably defined boundary and initial conditions on ρ. Let ρ_h be a numerically computed solution to the discrete analogue to (2.7). Then the residual is defined by

$$R_h = R_h(\rho_h) = \mathcal{D}(\rho_h) - f \text{ in } \mathcal{G}. \tag{2.8}$$

If we subtract the residual from both sides in (2.7) we obtain

$$\mathcal{D}(\rho - \rho_h) = f - R_h(\rho_h) \quad \Rightarrow \quad \varepsilon = \mathcal{D}^{-1}(f - R_h(\rho)), \qquad (2.9)$$

where $\varepsilon = \|\rho - \rho_h\|$ denotes the (discretization) error. One can easily see that the solution of (2.9) is at least as computationally expensive as the solution of the original problem (2.7). The trick is to find a method that solves (2.9) approximately and/or locally with much less effort than the original problem. $[\varepsilon]$ shall denote this approximate error estimator.

Another commonly used error estimation technique does not rely on the residual, but on the computed solutions themselves. Let ρ_h be the discrete solution of (2.7), while ρ_H is another solution obtained with a higher order scheme (if h and H represent the mesh size, we would required $h > H$). The Richardson extrapolation criterion defines $[\varepsilon]$ by

$$[\varepsilon] = \tilde{C} \|\rho_h - \rho_H\|, \qquad (2.10)$$

with $\tilde{C}$ a constant, correcting over- or under-estimation. In fact for $H \to 0$ we have that $[\varepsilon] \to \varepsilon$, provided that the method to compute ρ_H converges.

For an adaptive refinement control, we do not need to know the global error, but a local (element-wise) error. This can be achieved by translating the above idea to individual cells τ. Furthermore, we can define local area based residuals by

$$R_\tau(\rho_h) = \mathcal{D}(\rho_h) - f \text{ in } \tau,$$

and edge based (in 2D) or face based (in 3D) jumps, defined by

$$R_e(\rho_h) = \left| \frac{\partial \rho_h}{\partial n_e}|_{\tau_1} - \frac{\partial \rho_h}{\partial n_e}|_{\tau_2} \right|,$$

where τ_1 and τ_2 are the two cells interfacing at edge/face e, and n_e is the corresponding outer normal direction at edge e.

For finite element-like computations (i.e. for finite elements, as well as spectral elements and polynomial basis finite volume methods) we can distiguish five different basic approaches to derive a local a posteriori mathematical error estimator $[\varepsilon]_\tau$ (see also [71, 400, 401]).

1. **Residual estimators**: the local error is approximated by $[\varepsilon]_\tau$ by means of R_τ and R_e, defined above. This approach is due to Babuška and Rheinboldt [18]. Hugger has reviewed this approach [210] and Thomas and Sonar propose a residual estimator for nonlinear hyperbolic conservation laws [389].
2. **Estimators based on local Dirichlet problems**: For every element τ we solve a local problem of the form

$$\mathcal{D}(\chi) = f \text{ in } \overline{\tau}, \quad \chi = \rho_h \text{ on } \partial \overline{\tau}$$

where $\overline{\tau}$ is a slightly extended domain surrounding τ. The approximation order of this local problem is higher than the approximation order of ρ_h.

The estimator $[\varepsilon]_\tau$ is then derived from $\|\chi - \rho_h\|_{L^1(\tau)}$ Again this approach has been introduced by Babuška and Rheinboldt [17].

3. **Estimators based on local Neumann problems**: In this case a local problem of the form

$$\mathcal{D}(\chi) = R_\tau(\rho_h) \ in \ \tau, \quad \frac{\partial \chi}{\partial n} = R_e(\rho_h) \ on \ e \in \partial\tau,$$

is solved by an approximation order higher than the order of ρ_h. The error estimator is based on the energy norm $\|\chi\|_\tau$. These error estimators were originally proposed by Bank and Weiser [22]

4. **Estimators based on averaging**: Let us describe this method for Poisson's equation, i.e. $\mathcal{D} = -\Delta$. The method tries to construct an approximation $\sigma^{(2)}$ to the true $\Delta\rho$, by applying the following scheme for the gradient operator twice. Let σ be the weighted average of the gradients corresponding to one node patch:

$$\sigma = \sum_{\{i:\tau_i \in \ \text{patch of node} \ i\}} |\tau_i| \cdot \nabla\rho_h|_{\tau_i}.$$

As usual, $|\tau|$ denotes the area of τ. Extend σ to the whole element τ by linear interpolation. The error estimator is derived from $|\sigma^{(2)} - \Delta\rho_h|$. This approach can be found in [428]. Additionally, Carstensen shows that averaging methods are relyable and efficient [78].

5. **Hierarchical error estimators**: These estimators use an expanded finite element space and take the difference of both solutions by applying a strengthened Cauchy inequality. For the details see [122].

In atmospheric modeling the second approach can be found in [36]. An application of a variant of the averaging technique has been used in a mesh-less modeling technique in [41]. Furthermore, Skamarock derived a truncation error estimate following the lines of a Richardson extrapolation as outlined above [368]. Further reading in error estimation can include work by Bank and Xu [23, 24]. Recently, error estimation techniques based on computing error bounds have been proposed [86, 87, 272]. For hyperbolic problems, the error in an element is not only influenced by the local residual but also by the residual in the cells in the domain of influence. This dependency has been studied by Houston and Süli [206]. A proposal for efficient error estimates for computational fluid dynamics, based on hydrodynamic stability can be found in [226]. For biodegradation transport schemes Klöfkorn and coworkers propose an a posteriori error estimate that is suited for advection dominated problems [243].

We can conclude that there is still a large potential for improvement in finding good refinement criteria for atmospheric multi-component modeling. There is hardly any literature on combinations of error and refinement criteria in the presence of multiple refinement objectives. And a theoretically sound treatment of sub-grid processes in the error estimation is not available.

3

Grid Generation

We start the more formal description of adaptive atmospheric modeling methods with an introduction to mesh generation. In contrast to non-adaptive methods, where a fixed grid is given, in adaptive methods, the grid generating parts of a modeling software play a fundamental role. While in fixed grid applications, the software design is often derived from a computational stencil, in adaptive methods, the grid management forms the underlying basis for other derived data structures. We will discuss efficient data structures later in chap. 4 and will concentrate only on grid generation issues here.

Two different tasks characterize the art of mesh generation: Automatic generation of a (coarse) mesh for a given – in most cases polygonal – domain; and efficient refinement and un-refinement of a given initial mesh. Automatic mesh generation is still an open field of research, since up to now, there is no fully automatic meshing tool available that can cover all demands. However, some very impressive results have been gained and the reader is referred to the literature for an overview [229, 289, 391]. Since we are mostly concerned with spherical geometries without boundaries, we will focus on the second part, mesh refinement and un-refinement strategies, in this chapter. Section 3.5 reviews several initial grids for a special domain: the sphere.

3.1 Notation

In order to describe meshes formally correct, we need to introduce some notation. Let us begin with the basic definition of a triangulation. Note that by triangulation we do not necessarily mean a triangular mesh. So let us define a cell first, were we want to assume that cells in a triangulation are k-simplices.

Definition 3.1.1 *(k-simplex)*
   Let $\mathcal{P} = \{P_1, \ldots, P_k\} \subset \mathbb{R}^d$ be a (linearly independent) set of points (vertices) in the d-dimensional space. Then the convex hull of $\mathcal{P}$ is a k-simplex, where the convex hull is the intersection of all convex sets in $\mathbb{R}^d$ that contain $\mathcal{P}$.

Example 3.1.2 *(Simplicial cell types)*

- *A triangle in $\mathbb{R}^2$ is a 3-simplex.*
- *A hexahedral cell in $\mathbb{R}^3$ is a 8-simplex.*
- *A tetrahedral cell in $\mathbb{R}^3$ is a 4-simplex.*

Definition 3.1.3 *(Triangulation)*
   Let $\mathcal{G} \subsetneq \mathbb{R}^d$ be the finite d-dimensional and polygonal computational domain. Then $\mathcal{T} = \mathcal{T}(\mathcal{G}) = \{\tau_1, \ldots, \tau_M\}$, M number of cells, is an admissible triangulation, if

1. *the cells τ_i, $i = 1 : M$, are open disjoint k-simplices in $\mathcal{G}$, which cover the whole domain:*

$$\tau_i \cap \tau_j = \emptyset \quad \text{for } i \neq j; \quad \bigcup_{i=1:M} \overline{\tau_i} = \overline{\mathcal{G}}.$$

2. *for $i \neq j$ is $\overline{\tau_i} \cap \overline{\tau_j}$ either empty, or a common l-simplex with $l < k$ (a common edge, face, etc.).*

We used the following notation:

- *τ denotes the interior of a k-simplex, and*
- *$\overline{\tau}$ is the the closed k-simplex, i.e. its interior together with its boundary.*

Admissible triangulations are often called conforming *triangulations.*

Remark 3.1.4 *When considering triangular cells in $\mathbb{R}^2$, definition 3.1.3 is equivalent to the definition given in many graph-based textbooks (see [111]): Let $\mathcal{P} = \{P_1, \ldots, P_n\}$ be a set of points in the plane ($\mathbb{R}^2$). A maximal planar subdivision $\mathcal{S}$ is defined as a subdivision in which no edge can be added without destroying its planarity, i.e. any edge that is not in $\mathcal{S}$ intersects an edge in $\mathcal{S}$. An admissible triangulation is defined as a maximal planar subdivision with vertex set $\mathcal{P}$.*

Remark 3.1.5 *Note that definition 3.1.3 is cell based. Therefore, when dealing with node based methods, e.g. finite difference methods, only vertices of a triangulation are considered.*

Remark 3.1.6 *(Hanging node)*
   The above definition 3.1.3 defines an admissible triangulation, that is one without hanging nodes. A hanging node is a node on an unrefined edge (see fig. 3.1). Since many grid types incorporate hanging nodes we do not necessarily require admissible triangulations. A non-admissible triangulation is given by definition 3.1.3, omitting the second requirement. A non-admissible triangulation is often called non-conforming.

   In what follows, we will mostly omit the argument $\mathcal{G}$ for simplicity, since in most cases it will be clear or irrelevant how specifically $\mathcal{G}$ is defined. In order to characterize a triangulation we need a measure for the quality and an expression for hierarchies of triangulations.

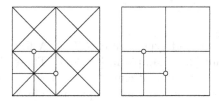

Fig. 3.1. Two hanging nodes each in two different triangulations

Definition 3.1.7 *(Hierarchies of triangulations)*
   Let $T_1 = \{\tau_1, \ldots, \tau_M\}$ and $T_2 = \{\tau_1, \ldots, \tau_K\}$ two triangulations. If T_2 is created from T_1 by refinement, we write $T_1 \prec T_2$.

Definition 3.1.8 *(Mesh width, inner angle and regularity)*
   Let the following parameters be defined for a cell τ in a triangulation T.

$$h_{e,\tau} : \text{length of longest edge in } \tau,$$
$$h_{B,\tau} : \text{radius of circum-circle of } \tau,$$
$$h_{b,\tau} : \text{radius of inner circle of } \tau.$$

We define the following parameters for characterizing the mesh size of T:

$$h_e = \min_{\tau \in T}\{h_{e,\tau}\},$$
$$h_B = \min_{\tau \in T}\{h_{B,\tau}\},$$
$$h_b = \min_{\tau \in T}\{h_{b,\tau}\}.$$

Let furthermore ϑ_τ be the smallest angle in τ. The smallest inner angle of T is given by

$$\vartheta = \min_{\tau \in T}\{\vartheta_\tau\}.$$

Note that for $\mathbb{R}^d$ with $d > 2$ the inner angle definition has to be replaced by an inner cone. Finally, T is called regular, *if $0 \ll \vartheta \ll \pi$.*

   Note that since
$$h_{b,\tau} < h_{e,\tau} < h_{B,\tau} < C_\vartheta h_{b,\tau},$$
where C_ϑ is a constant that depends only on the smallest inner angle, all mesh width parameters are equivalent to each other. Therefore, we often just use the notation h for mesh width, without specifying which specific measure is to be used.

Remark 3.1.9 *Since the sum of all inner angles of a triangle is π, for triangulations with triangular elements, the definition of* regular *in definition 3.1.8 implies that ϑ is close to $\frac{\pi}{3}$. For quadrilateral triangulations a similar remark holds with ϑ close to $\frac{\pi}{2}$.*
   Note that the regularity condition is equivalent to $0 \ll \frac{h_b}{h_B} = C_\vartheta^{-1} < 1$.

3.2 Grid Types

In this section different grid types are classified. A first attempt to classify grid types can be found for example in [88, 204]. These classifications have been developed mainly for automatic mesh generation methods in engineering. However, since we are concerned with adaptively refined grids, we propose a distinct approach here.

The first criterion of distinction between grids is the admissibility condition. Admissible (or conforming) grids do not allow hanging nodes. Both quadrilateral and triangular grids can have hanging nodes, when refined locally, depending on the refinement strategy.

A second important distinction between grids is their orthogonality. Quadrilateral grids have orthogonality properties, making them suitable for using 1D basis functions with tensor product expansion to d dimensions. When considering higher dimensional triangulations, also mixed grid types can occur, where the horizontal grid is (non-orthogonal) triangular while the vertical dimension is discretized by an orthogonal quadrilateral extension, to give an example.

An important distinction is nesting. For locally refined meshes non-nested grids are created by defining new nodes and then completely re-meshing the triangulation. With nested grids, local refinement of single mesh elements with a certain refinement strategy is performed. Nested meshes allow for (more or less) straight forward implementation of multi-grid methods. Note that definition 3.1.7 is more general, since it does not detail the refinement strategy.

A fourth category of distinction concerns structure, which we formally define.

Definition 3.2.1 *(Structured and unstructured mesh)*
A mesh is called structured, *if the nodes can be ordered by an (incrementing) index array of d dimensions* $(i_1, i_2, \ldots, i_d)$ *and if neighboring nodes can be accessed by incrementing/decrementing the index, e.g.* $(i_1, i_2 + 1, \ldots, i_d)$. *Otherwise a mesh is called* unstructured.

Note that for an unstructured mesh additional data is required to define neighborhood relations (this is the so called connectivity). In adaptive grid generation, even logically structured meshes are sometimes treated like unstructured meshes, because then no different approaches have to be used in uniform (structured) and locally refined (unstructured) areas.

For a short and precise descriptions of meshes we introduce a naming convention for grid types given in table 3.1. The first four positions of the name-code correspond to the above classification, while the fifth position corresponds to a name for the geometry of grid cells. In case a grid does not have one or the other of the tabulated characteristic properties, the naming code is omitted. With this naming convention, a locally refined block-structured quadrilateral grid as used e.g. in [221] could be called a ONSH-grid, while a

Table 3.1. Naming convention for adaptive grids

Position	Tag	Description
1	C	C: conforming
2	O	O: orthogonal
3	N	N: nested
4	S	S: structured
5	T/H/P	T: triangular/tetrahedral
		H: quadrilateral/hexahedral
		P: polygonal/polyhedric

bisected triangular grid like in amatos would be called CNT-grid. A Delaunay triangulation is of type CT.

In order to further characterize a mesh, we can look at the number of refinements as an indicator of resolution ratio. Most gridding schemes start with a given initial mesh, which is uniformly refined up to a given level of refinement and then further refined locally. For triangular meshes derived from an icosahedral initial discretization of the sphere, Baumgardner and Frederickson and more recently Giraldo have introduced a classification of uniform triangular refinements [167, 30]. For a more universal convention and assuming that we have nested grids, we propose a pair of numbers $[c : f]$ where c indicates the number of uniform refinements, i.e. c represents the coarsest mesh level used in computations, and f is the finest grid level.

Extending the above naming convetion with this pair, we have a ONSH[3 : 6]-grid in [221], while in [39] the finest mesh is of type CNT[4 : 19]. If we have uniform grids, then we can omit one of the entries, obtaining CNT[15] $\equiv$ CNT[15 : 15].

If refinement levels are inadequate or not available for the classification, we could consider resolutions of the coarsest grid cells and the finest grid cells. Thus, CNT[4 : 19] $\equiv$ CNT$(230 : 5)_{[km]}$. For distinction, we have put the resolution in brackets instead of square brackets and added a subscript denoting the unit (Kilometers here). As another example, the operational global model (GME) of the German Weather Service (DWD) has type CSH$(40)_{[km]}$.

3.3 Refinement Strategies in 2D

In this section we give an overview of commonly used refinement strategies. Selection of a strategy depends on the application and the objectives of meshing. One could – for example – either optimize grid quality, but faces algorithmic complexity. Or it could be important to achieve orthogonality, but then admissibility cannot be maintained. There is no single strategy that optimizes all of the existing desirable properties of a mesh. For a good overview of mesh generation especially in engineering applications, see [391]. A comparison of refinement strategies in the context of finite element methods is given in [290].

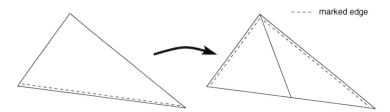

Fig. 3.2. Bisection of marked edge algorithm 3.3.1 acting on a triangle

Triangular Bisection

A very simple, yet powerful refinement strategy for triangular meshes is the bisection strategy. It constructs two daughter triangles from one mother. In order to maintain regularity of the mesh, only marked edges are allowed to be refined. Defined in an iterative procedure, this strategy leads to locally adaptive, nested, and admissible meshes.

Algorithm 3.3.1 *(Bisection of marked edge)*
Let τ be a triangle, defined by its vertices v_1, v_2, v_3. Without loss of generality let $e_m = \overline{\{v_1, v_2\}}$ be a marked edge.

1. Insert new vertex v_n at center of e_m, i.e. $v_n = (v_1 + v_2)/2$.
2. Define new cells τ_1 and τ_2 by:

$$\tau_1 = \{v_1, v_n, v_3\}, \quad \tau_2 = \{v_n, v_2, v_3\}.$$

3. Mark unrefined edges of τ in elements τ_i, i.e. $e_m^{\tau_1} = \overline{\{v_1, v_3\}}$, and $e_m^{\tau_2} = \overline{\{v_2, v_3\}}$.

This algorithm is depicted in fig. 3.2. Note that the order of vertices is important to maintain the orientation of triangles. The above definition maintains orientation in the daughter elements.

Remark 3.3.2 *(Bisection of longest edge)*
There is a variant of algorithm 3.3.1, called bisection of longest edge. It differs from the above given algorithm in that it refines the longest edge only. When starting with a sufficiently regular initial triangle, both algorithms are equivalent. However, marking an edge with the above scheme results in less computational effort, because longest edge bisection needs to evaluate the edge length for each of the three edges in each step.

In order to create a complete mesh with regions of local refinement we have to use the bisection algorithm iteratively:

Algorithm 3.3.3 *(Admissible bisection triangulation)*
Let $\mathcal{T} = \{\tau_1, \ldots, \tau_M\}$ be a given admissible coarse triangulation, and let $\mathcal{S} \subset \mathcal{T}$ be the set of cells marked for refinement.
WHILE $(\mathcal{S} \neq \emptyset)$

Fig. 3.3. The iterative process to construct a triangulation by algorithm 3.3.3. The shaded area corresponds to the set $\mathcal{S}$, dotted lines depict marked edges, and a hanging node is marked by an open circle

1. **FOR EACH** *($\tau \in \mathcal{S}$): refine τ according to algorithm 3.3.1, and obtain a new triangulation $\mathcal{T}^*$, remove τ from $\mathcal{S}$ ($\mathcal{S}$ is empty at the end of this step);*
2. **FOR EACH** *($\tau \in \mathcal{T}^*$):* **IF** *τ has hanging node* **THEN**: $\mathcal{S} = \mathcal{S} \cup \{\tau\}$.

END WHILE

Algorithm 3.3.3 is visualized in fig. 3.3. It can be shown that algorithm 3.3.3 converges, i.e. $\mathcal{S}$ is in fact empty after a finite number of iterations. Furthermore, it leads to an admissible triangulation and – except for single pathological cases – local refinement has no global effect on the mesh (see [25, 196, 341]).

This algorithm is simple from a coding point of view. It leads to admissible triangulations and maintains regularity. In fact, the inner angles are bounded by $\frac{1}{2}\vartheta_0$, where ϑ_0 is the smallest inner angle of the initial mesh. Triangulations of different levels are nested. Each level decreases the mesh size by a factor of $\sqrt{2}$. An example of application of this algorithm can be found in [39].

Definition 3.3.4 *(Admissible patch)*
An admissible patch is a grid part consisting of four triangles grouped around a refined edge (see fig. 3.4).

Remark 3.3.5 *(Coarsening bisected triangulations)*
In order to coarsen a bisected triangulation, we look for admissible patches. If at least three elements of an admissible patch are flagged for coarsening, then all four elements are deleted, obtaining a new mesh section with only the two mother triangles.

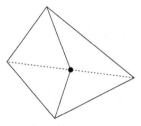

Fig. 3.4. Admissible patch with refinement node •, and the refinement edge (dashed)

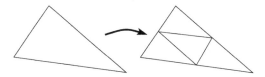

Fig. 3.5. Regular refinement algorithm 3.3.6 acting on a triangle

Regular Triangular Refinement

Regular refinement of triangles divides each edge of a given initial triangle:

Algorithm 3.3.6 *(Regular triangular refinement)*
   Given a triangle τ, defined by its vertices v_1, v_2, v_3 with edges $e_1 = \overline{\{v_2, v_3\}}$, $e_2 = \{v_1, v_3\}$, $e_3 = \{v_1, v_2\}$.
Insert new vertices $v_{n,i}$ at center of e_i, e.g. $v_{n,1} = (v_2 + v_3)/2$.
Define new cells $\tau_1, \dots, \tau_4$ by:

$$\tau_1 = \{v_1, v_{n,3}, v_{n,2}\}, \qquad \tau_2 = \{v_{n,3}, v_2, v_{n,1}\}$$
$$\tau_3 = \{v_{n,2}, v_{n,1}, v_3\}, \qquad \tau_4 = \{v_{n,1}, v_{n,2}, v_{n,3}\}$$

   The algorithm is depicted in fig. 3.5. This refinement strategy leads to non-admissible meshes. Therefore, if admissibility is required, we have to discuss methods to achieve this goal. As in the previous section, the sequence of vertices given in the algorithm above preserves orientation.
   Most authors introduce a *green refinement* for admissibility. This is a temporary bisection refinement of those cells that have one hanging node. With green refinement the hanging node and its opposite vertex are connected to build an edge, dividing the cell into two daughter cells (see figure 3.6). We now formulate an algorithm that is commonly used to create admissible triangulations (see [34] for an example).

Algorithm 3.3.7 *(Admissible regularly refined triangulation)*
   Let $T = \{\tau_1, \dots, \tau_M\}$ be a given admissible coarse triangulation, and let $S \subset T$ be the set of cells marked for regular refinement.

1. Remove all green closures from previous refinement step.
2. WHILE $(S \neq \emptyset)$

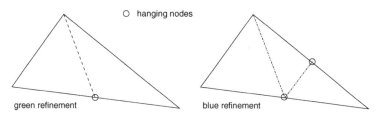

Fig. 3.6. Green and blue refinement to prevent hanging nodes

a) FOR EACH $(\tau \in \mathcal{S})$: refine τ according to algorithm 3.3.6, and obtain a new triangulation $\mathcal{T}^{\star}$, remove τ from $\mathcal{S}$;

b) FOR EACH $(\tau \in \mathcal{T}^{\star})$: IF τ has more than one hanging node, THEN: $\mathcal{S} = \mathcal{S} \cup \{\tau\}$.

END WHILE.

3. FOR EACH $(\tau \in \mathcal{T}^{\star})$: IF τ has hanging node, THEN: refine τ by green refinement.

Remark 3.3.8 *As in subsect. 3.3, it can be shown that algorithm 3.3.7 leads to an admissible triangulation. The inner angles are bounded, since regular refinement creates four similar daughter triangles form each mother triangle.*

Remark 3.3.9 *Some authors introduce a* blue refinement *to treat cells with two hanging nodes. In order to obtain a blue refinement, proceed as follows (see fig. 3.6):*

1. *bisect the cell by taking the hanging node on the longer edge and its opposite vertex as bisection edge;*
2. *bisect the daughter element of step 1, which contains the remaining hanging node by connecting this hanging node with the first hanging node.*

Remark 3.3.10 *(Coarsening regularly refined triangulation)*
Since green or blue refinements are removed from the grid before refinement/coarsening, we only have to consider regularly refined elements for coarsening. So, in order to coarsen a regularly refined triangular element, the four daughter elements are removed, leaving the mother on the now finest mesh level. Coarsening is performed if three or more daughters are flagged for coarsening.

Quadrilateral Refinement

Quadrilateral grids are very popular since several advantages can be exploited for efficient numerical calculations. Quadrilaterals are well suited for efficient interpolation/quadrature schemes, because the local coordinate system can be formulated orthogonally. A structured data layout can be achieved easily. And the ratio of cell number to node number (which is a measure for computational efficiency) is smaller than in triangular meshes. On the other hand, admissibility is often hurt. And more importantly, complex geometries are not easily represented by quadrilaterals. As a simplification and requirement for orthogonality, we want to assume convex quadrilaterals throughout this section.

The most simple refinement strategy, is to insert one vertex and add edges to the centers of all cell edges:

Algorithm 3.3.11 *(General quadrilateral refinement)*
Given a Quad χ, defined by its vertices $v_1, \ldots, v_4$ with edges $e_1 = \overline{\{v_1, v_2\}}$, $\ldots$, $e_4 = \overline{\{v_4, v_1\}}$.

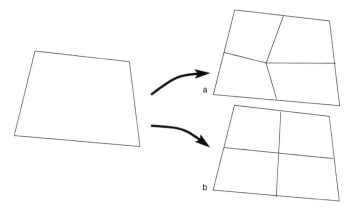

Fig. 3.7. Algorithm 3.3.11 for general quadrilateral refinement (a) and 3.3.12 for regular quadrilateral refinement acting on a cell

1. *Insert new vertex v_n inside χ.*
2. *Create new edge vertices $v_{n,1}, \ldots, v_{n,4}$ by*

$$v_{n,1} = \frac{(v_1 + v_2)}{2}, \ldots, v_{n,4} = \frac{(v_4 + v_1)}{2}.$$

3. *Define new cells $\chi_1, \ldots, \chi_4$ by:*

$$\chi_1 = \{v_1, v_{n,1}, v_n, v_{n,4}\}, \quad \chi_2 = \{v_{n,1}, v_2, v_{n,2}, v_n\}$$
$$\chi_3 = \{v_n, v_{n,2}, v_3, v_{n,3}\}, \quad \chi_4 = \{v_{n,4}, v_n, v_{n,3}, v_4\}$$

Remark 3.3.12 *(Regular quadrilateral refinement)*

A less general quadrilateral refinement is given by a variant of algorithm 3.3.11, where we omit step 1 and define the newly inserted vertex as the intersection of the new edges $e_{n,1} = \overline{\{v_{n,1}, v_{n,3}\}}$ and $e_{n,2} = \overline{\{v_{n,2}, v_{n,4}\}}$. Both variants are depicted in fig. 3.7.

Remark 3.3.13 *(Closure for general quadrilateral refinement)*

Quadrilateral refinement leads to hanging nodes in the triangulation so we have to deal with closures. A common closure uses triangular cells to remove hanging nodes. In this method we have to differentiate five different cases (see fig. 3.8 for the first four cases):

1. **One hanging node:** *connect hanging node with both vertices of opposite edge, obtaining 3 new triangular cells;*
2. **two hanging nodes on opposite edges:** *connect both hanging nodes, obtaining 2 new quadrilateral cells;*
3. **two hanging nodes on adjacent edges:** *insert new vertex at cell center, connect new vertex with hanging nodes and opposite cell vertex, obtaining 3 new quadrilateral cells;*

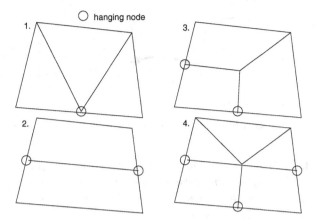

Fig. 3.8. Closures for avoiding hanging nodes in quadrilateral grid refinement. Four cases are shown, as described in the text

4. **three hanging nodes:** *insert new vertex at cell center, connect new vertex nodes with all hanging, and with vertices of unrefined edge, obtaining 2 new quadrilateral and 3 new triangular cells;*
5. **four hanging nodes:** *refine regularly.*

Most of the current quadrilateral grid management packages treat no single cells but patches of cells. Thus, let the patch size be $\nu \times \nu$, then refinement consists of creating a new $\nu \times \nu$-patch in each cell marked for refinement.

Algorithm 3.3.14 *(Quadrilateral 3×3-patch refinement)*
Let $v_1, \ldots, v_4$ be the vertices of a quadrilateral cell χ in counter-clockwise order starting from the bottom left vertex. Then denote by index i the local x-direction and by index j the local y-direction correspondingly. We want to denote the new vertices by $v_{i,j}$. Note that for a 3×3-patch $i, j = 0 : 3$, with v_1 corresponding to index pair $(0,0)$, i.e. $v_1 = v_{0,0}$, $v_2 = v_{3,0}$, $v_3 = v_{3,3}$, $v_4 = v_{0,3}$.

1. *Insert new vertices $v_{i,j}$, $i, j = 0 : 3$ using the following formula:*

$$v_{i,j}(x) = \frac{1}{9} \left((3 - j)\left[(3 - i)v_1(x) + iv_2(x)\right] + j\left[(3 - i)v_4(x) + iv_3(x)\right]\right);$$
$$v_{i,j}(y) = \frac{1}{9} \left((3 - i)\left[(3 - j)v_1(y) + jv_4(y)\right] + i\left[(3 - j)v_2(y) + jv_3(y)\right]\right);$$

   where $v_{i,j}(x)$ denotes the x-component of $v_{i,j}$, and $v_{i,j}(y)$ the y-component respectively.
2. *Now, the nine new cells of the refined 3×3-patch are given by $\{v_{i,j}, v_{i+1,j}, v_{i+1,j+1}, v_{i,j+1}\}$ with $i, j = 0 : 2$.*

Figure 3.9 shows the action of algorithm 3.3.14.

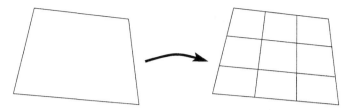

Fig. 3.9. Action of algorithm 3.3.14 on a quadrilateral cell, marked for refinement

Remark 3.3.15 *(Closures for quadrilateral 3×3-patch refinement)*

Again, in order to find admissible triangulations, a closure has to be found. In the 3×3-patch case, four situations are considered (see fig. 3.10 for the first two cases):

1. **One edge has hanging nodes:** *Without loss of generality, let $v_{1,0}$, and $v_{2,0}$ be the hanging nodes. Then insert $v_{1,1}$, and $v_{2,1}$. The four closing quadrilateral sub-cells are given by*

$$\{v_{0,0}, v_{1,0}, v_{1,1}, v_{0,3}\}, \quad \{v_{1,0}, v_{2,0}, v_{2,1}, v_{1,1}\},$$
$$\{v_{2,0}, v_{3,0}, v_{3,3}, v_{2,1}\}, \quad \{v_{1,1}, v_{2,1}, v_{3,3}, v_{0,3}\}.$$

2. **Two adjacent edges have hanging nodes:** *Without loss of generality, let $v_{1,0}$, $v_{2,0}$, $v_{0,1}$, and $v_{0,2}$ be the hanging nodes. Then insert $v_{1,1}$ and $v_{2,2}$. The five closing quadrilateral cells are given by*

$$\{v_{0,0}, v_{1,0}, v_{1,1}, v_{0,1}\},$$
$$\{v_{1,0}, v_{2,0}, v_{2,2}, v_{1,1}\}, \quad \{v_{2,0}, v_{3,0}, v_{3,3}, v_{2,2}\},$$
$$\{v_{0,1}, v_{1,1}, v_{2,2}, v_{0,2}\}, \quad \{v_{0,2}, v_{2,2}, v_{3,3}, v_{0,3}\}.$$

3. **Two opposite edges have hanging nodes:** *In this case, just connect the corresponding opposing hanging nodes, obtaining three closing quadrilaterals.*

4. **Three or four edges have hanging nodes:** *refine regularly.*

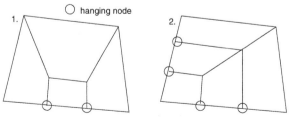

Fig. 3.10. Closures for avoiding hanging nodes in quadrilateral patch refinement. Two cases are shown, as described in the text

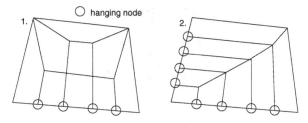

Fig. 3.11. Generalization of closures for 5×5-patch

Remark 3.3.16 *For uneven numbers of sub-cells in each patch cell, a convenient closure method can be derived that avoids triangular closures. We gave an example for 3×3-patches in algorithm 3.3.15 that can easily be generalized to higher uneven numbers. It should be mentioned, however, that for high numbers $\nu = 2k + 1$, $k \in \mathbb{N}$ very thin quadrilaterals can occur which might degrade numerical stability. See fig. 3.11 for an example with a 5×5-patch.*

Finally, in order to obtain an algorithm that creates an admissible mesh, the refinement and closure strategies have to be combined. This algorithm has been used in [154].

Algorithm 3.3.17 *(Admissible quadrilateral triangulation)*
    Let $\mathcal{Q} = \{\chi_1, \ldots, \chi_M\}$ be a given admissible coarse quadrilateral triangulation, and let $\mathcal{S} \subset \mathcal{Q}$ be the set of cells marked for regular refinement.

1. *Remove all closures from previous refinement step.*
2. WHILE *($\mathcal{S} \neq \emptyset$)*
    a) FOR EACH *($\chi \in \mathcal{S}$): refine χ according to algorithm 3.3.11 or 3.3.14, and obtain a new triangulation $\mathcal{Q}^\star$, remove χ from $\mathcal{S}$;*
    b) FOR EACH *($\chi \in \mathcal{Q}^\star$):* IF *$\chi$ has more than 2 resp. 3 edges with hanging nodes,* THEN: *$\mathcal{S} = \mathcal{S} \cup \{\chi\}$.*
    END WHILE.
3. FOR EACH *($\chi \in \mathcal{Q}^\star$):* IF *$\chi$ has hanging nodes,* THEN: *refine χ using remarks 3.3.13 and 3.3.15, resp.*

Remark 3.3.18 *Again, it can be shown that algorithm 3.3.17 leads to an admissible quadrilateral triangulation. Inner angles in quadrilateral triangulations are obviously bounded, provided that the initial triangulation is well behaved.*

Delaunay Triangulation – Non-Nested Refinement

For the definition of a Delaunay triangulation, we start with the definition for the *Voronoi Diagram* of a point set $\mathcal{P} = \{v_1, \ldots, v_n\}$. As in the previous sections, we will denote the points by v_i, since we also call them vertices.

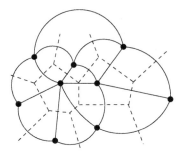

Fig. 3.12. The Voronoi cells (*dotted lines*) of a point set $\mathcal{P}$ and the Delaunay graph (*solid lines*)

Definition 3.3.19 *(Voronoi Diagram)*
  Let $\mathcal{P} = \{v_1, \ldots, v_n\} \subset \mathbb{R}^2$ be a given set of points in the plane. Then the Voronoi Diagram of $\mathcal{P}$ is the subdivision of the plane into n sub-domains V_i corresponding to the n vertices v_i, $i = 1 : n$, such that

$$V_i = \{x \in \mathbb{R}^2 : \text{dist}(x, v_i) \leq \text{dist}(x, v_j),\ j \neq i\},$$

where $\text{dist}(x, y)$ is the Euclidean distance between x and y. The sub-domain V_i is called Voronoi cell corresponding to vertex v_i.

  Now, we proceed with the *Delaunay Graph* of a point set $\mathcal{P}$ (see fig. 3.12 for a Voronoi diagram together with its Delaunay graph).

Definition 3.3.20 *(Delaunay Graph)*
  As above, Let $\mathcal{P}$ be a set of points/vertices, and let $\mathcal{V} = \{V_1, \ldots, V_n\}$ be the Voronoi diagram. Then the Delaunay graph is defined by the graph nodes $\{v_1, \ldots, v_n\}$ and connecting arcs for each pair (v_i, v_j), $i \neq j$, wherever the Voronoi cells V_i and V_j share an edge.

  Finally, connecting all graph-connected vertices by straight edges, induces the *Delaunay Triangulation* (see fig. 3.13). Neglecting some pathological cases, it can be proven that in deed, this procedure yields an admissible triangulation in the sense of defining remark 3.1.4. Furthermore, it can be shown (see e.g. [111]) that the Delaunay triangulation of a point set $\mathcal{P}$ is maximizing the inner angles of the Triangulation. Or in other words, all interior angles are bounded away from $0°$ and $90°$, resp.

  The Delaunay triangulation of a point set $\mathcal{P}$ can be computed in $\mathcal{O}(n \log n)$ operations with $\mathcal{O}(n)$ required storage positions, where n is the number of vertices in $\mathcal{P}$. A practical algorithm (also for 3D Delaunay trianglulations) is given in [383].

  In order to construct an adaptive method using a Delaunay triangulation, we need to specify a vertex insertion method. Again, we assume that there is a given initial triangulation with a set of triangles marked for refinement. In principle, one could take a random distribution of points as an initial triangulation.

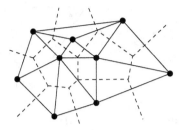

Fig. 3.13. The Delaunay triangulation (*solid lines*) of a given point set together with the Voronoi diagram (*dotted lines*)

Algorithm 3.3.21 *(Unstructured Vertex insertion)*
    Given a triangle τ, defined by its vertices v_1, v_2, v_3.
Insert a new vertex v_n at the barycenter of τ: $v_n = (v_1 + v_2 + v_3)/3$.
Define new cells $\tau_1, \ldots, \tau_3$ by:

$$\tau_1 = \{v_1, v_n, v_2\}, \tau_2 = \{v_2, v_n, v_3\}, \tau_3 = \{v_3, v_n, v_1\}$$

This algorithm is used in [20]. The insertion of new cells is depicted in fig. 3.14.

Remark 3.3.22 *Note that algorithm 3.3.21 does not prevent inner angles to collapse. Therefore, in the algorithm for creating an complete triangulation, a re-meshing step is required (see algorithm 3.3.24, resp. remark 3.3.25).*

Remark 3.3.23 *Another vertex-based possibility to find locations for new nodes has been used in [41]. Here, vertices are flagged for refinement instead of cells. Then, each corner of the corresponding Voronoi-cell is inserted as new vertex. After insertion of new nodes, re-meshing has to be performed. This method is especially suited for mesh-free methods, where only point locations are considered for calculations.*

The complete algorithm for obtaining an admissible locally refined triangulation from a given Delaunay triangulation follows:

Algorithm 3.3.24 *(Admissible unstructured triangulation)*
    Let $T = \{\tau_1, \ldots, \tau_M\}$ be a given admissible coarse triangulation, and let $S \subset T$ be the set of cells marked for refinement.

Fig. 3.14. Local refinement of a Delaunay triangulation by insertion of centroid and re-meshing

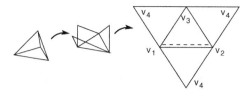

Fig. 3.15. Unfolding a tetrahedron into standard position, and notations for local vertex ordering and globally marked edge (*dotted line*)

1. FOR EACH $(\tau \in \mathcal{S})$: *refine τ according to algorithm 3.3.21;*
2. FOR EACH $(\tau \in \mathcal{S})$:
   a) FOR EACH *edge e_i, $(i = 1 : 3)$:*
      IF *e_i is not a boundary edge, then if edge flipping yields larger minimum inner angle, flip edges;*
   b) *remove τ form $\mathcal{S}$.*

Remark 3.3.25 *Note that step 2 of the above algorithm could be exchanged by the following instruction:*

2a Create Delaunay triangulation of new point set, obtained by taking all vertices after insertion.

The last algorithm was used in [166], but it is often not an efficient way of adaptive computation.

3.4 Refinement in 3D

In this section, we will extend the methods described in the previous section to three dimensional space $\mathbb{R}^3$. This will not be done in such detail, since most of the extensions presented are rather straight forward and a complete technical description would render this section somewhat tedious and lengthy. We will concentrate on the 3D-extension of the triangle bisection (algorithm 3.3.1) and the regular quadrilateral refinement (algorithm 3.3.11) together with its closure (algorithm 3.3.13).

Tetrahedral Bisection

In this section we follow an approach introduced by Bänsch [25]. We start with defining a tetrahedron with marked edges.

Definition 3.4.1 *(Tetrahedron with globally marked edge)*
    Let τ^{3D} be a tetrahedron, defined by its four vertices $\{v_1, \ldots, v_4\} \subset \mathbb{R}^3$, with faces $\tau_1 = \overline{(v_1, v_2, v_3)}$, $\tau_2 = \overline{(v_2, v_4, v_3)}$, $\tau_3 = \overline{(v_1, v_3, v_4)}$, and $\tau_4 = \overline{(v_1, v_4, v_2)}$ (see fig. 3.15. Let each of the τ_i have one marked edge such that exactly two of the faces share their marked edge. We will call this edge a globally marked edge (of tetrahedron τ^{3D}).

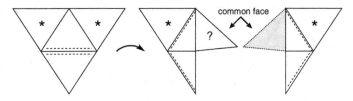

Fig. 3.16. Bisection of a tetrahedron (action of algorithm 3.4.3)

Remark 3.4.2 *Note that the representation in fig. 3.15 is called* standard position. *After rotation and scaling, this representation is unique, assuming that there is only one globally marked edge* $\overline{(v_1, v_2)}$.

When refining a tetrahedron, we can simply apply the 2D algorithms for refining triangles to the faces. However, we have to define, how to set the marked edge in the newly created face separating the two new tetrahedra (see fig. 3.16). Bänsch [25] uses the notion of red and black tetrahedra as given in fig. 3.17.

Algorithm 3.4.3 *(Tetrahedral Bisection)*
   Given a tetrahedron τ^{3D} *in standard position as defined in definition 3.4.1, with a globally marked edge* $e_m = \overline{(v_1, v_2)}$. *We will denote the mother of* τ^{3D} *by* $\mathcal{M}(\tau^{3D})$.

1. *Insert new vertex* v_n *at center of* e_m.
2. *Refine both faces sharing* e_m *according to algorithm 3.3.1 for 2D bisection.*
3. *Insert new face with vertices* $\{v_n, v_4, v_3\}$ *and define the marked edge* e_m^{new} *according to the following scheme:*
   a) *IF* $(\tau^{3D}$ *is black and* $\mathcal{M}(\tau^{3D})$ *is red):* $e_m^{\text{new}} = \overline{(v_3, v_4)}$.
   b) *IF* $(\tau^{3D}$ *is red):* $e_m^{\text{new}} = \overline{(v_3, v_4)}$.
   c) *IF* $(\tau^{3D}$ *is light black and* $\mathcal{M}(\tau^{3D})$ *is black):* $e_m^{\text{new}} = \overline{(v_n, v_4)}$.
   d) *IF* $(\tau^{3D}$ *is dark black and* $\mathcal{M}(\tau^{3D})$ *is black):* $e_m^{\text{new}} = \overline{(v_n, v_3)}$.
   e) *ELSE:* $e_m^{\text{new}} = \overline{(v_3, v_4)}$.

Now, using this refinement scheme, the algorithm for obtaining an admissible 3D triangulation (or rather tetrangulation) proceeds exactly as in the 2D case (see algorithm 3.3.3). Bänsch showed for the 3D case the same termination property as for the 2D algorithm [25]

Algorithm 3.4.4 *(Admissible 3D bisection triangulation)*
   Let $\mathcal{T} = \{\tau_1^{3D}, \dots, \tau_M^{3D}\}$ *be a given admissible coarse triangulation, and let* $\mathcal{S} \subset \mathcal{T}$ *be the set of cells (tetrahedra) marked for refinement.*
WHILE $(\mathcal{S} \neq \emptyset)$

1. FOR EACH $(\tau^{3D} \in \mathcal{S})$: *refine* τ^{3D} *according to algorithm 3.4.3, and obtain a new triangulation* $\mathcal{T}^*$, *remove* τ^{3D} *from* $\mathcal{S}$;
2. FOR EACH $(\tau^{3D} \in \mathcal{T}^*)$: IF τ^{3D} *has hanging node* THEN: $\mathcal{S} = \mathcal{S} \cup \{\tau^{3D}\}$.

END WHILE

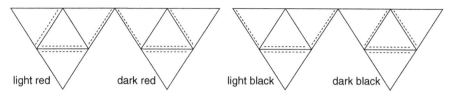

Fig. 3.17. Red and black tetrahedral layout in standard position

Hexahedral Refinement

As for the extension of 2D triangular refinement strategies to 3D, the hexahedral refinement is just an extension of the 2D quadrilateral refinement given in algorithm 3.3.12. But instead of using arbitrary vertex numbers v_n, as in the 2D case, we introduce local coordinates, taking advantage of the virtually orthogonal structure of quadrilateral grids. The lower left front corner has coordinates $[0, 0, 0]$, where $[\cdot, \cdot, \cdot]$ denotes the local coordinate system, while the upper right back corner has coordinates $[1, 1, 1]$ or $[2, 2, 2]$ after re-normalization and refinement (see fig. 3.18).

Algorithm 3.4.5 *(Regular hexahedral refinement)*
    Given a hexahedron τ^{3D}, defined by its vertices $v_1 = [0, 0, 0], \ldots, v_8 = [1, 1, 1]$, with faces $\nu_1, \ldots, \nu_6$.

1. *Refine each face ν_i $(i = 1 : 6)$, using algorithm 3.3.12.*
2. *Connect opposite center vertices. The intersection of these edges yields another new vertex v_n.*
3. *Define new cells $\tau_{0,0,0}^{3D}, \ldots, \tau_{1,1,1}^{3D}$ by:*

$$\tau_{i,j,k}^{3D} = \{ \ [0, 0, 0], [1, 0, 0], [1, 1, 0], [0, 1, 0] \ ,$$
$$[0, 0, 1], [1, 0, 1], [1, 1, 1], [0, 1, 1] \ \} + [i, j, k], \quad i, j, k = 0, 1.$$

As in the 2D case, hexahedral refinement causes non-admissible triangulations, i.e. hanging nodes. Therefore, a closure mechanism has to be defined, if admissible triangulations are required. For an extension of the 2D algorithm, pyramidal elements can be introduced. However, the number of cases that have to be tackled and the region of influence for closures is by far not as simple as in the 2D case. In fact, it is not possible to define a point insertion algorithm that results in a conforming (purely) hexahedral mesh in complete generality (see chap. 21 in [391] by R. Schneiders). So we will omit the discussion of all those cases here.

Most applications with hexahedral mesh refinement use solution techniques that can handle hanging nodes. Thus, a non-conforming block-wise refinement strategy for local refinement is often applied (see for example [207]).

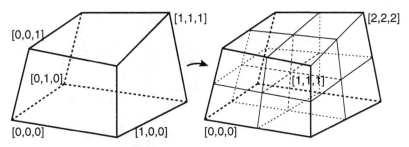

Fig. 3.18. Hexahedral refinement, the action of algorithm 3.4.5

3.5 Spherical Geometries

In order to find discrete representations of the sphere, we have to start with an initial triangulation $\mathcal{T}$. In most cases, the sphere is triangulated by a 2D triangulation, which is extended to fully 3D geometries by layers of identical 2D grids.

Triangular Initial Triangulations

In this subsection we will introduce three different possibilities to triangulate a polygonal approximation to the sphere. In order to represent a spherical domain with triangular elements, we need to either define spherical triangles, or agree to be satisfied with a polygonal representation of the sphere.

When considering the triangulation as a basis for further numerical calculations, one has to keep in mind that a polygonal representation with plane cells introduces first order error terms. Thus, combining a plane cell type with a high order method in spherical geometries is a waste of computational effort. The mathematically correct triangulation which preserves the order, would use curved parametric elements. An introduction to this type of elements can be found in [359]. In order to maintain the order of accuracy it is possible to use a plane polygonal representation but to lift the high order sample points within each element to positions on the sphere's surface. We will proceed with this approach, and can now start to introduce basic polygonal representations of the sphere.

A natural starting point for spherical triangulations are the platonic solids. Since we are interested in triangular elements in this section, we do not consider the cube in first place. The next choice could be the tetrahedron, but it does not represent a sphere too well. A tetrahedron would be better suited for a hemi-sphere. The next simplest polygonal representations of the sphere is an octahedron with six vertices, 12 edges and eight triangular faces.

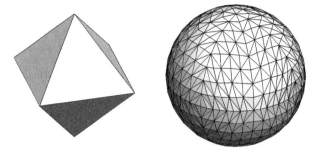

Fig. 3.19. Octahedral triangulation of sphere: (**left**) initial triangulation, with 8 triangles; (**right**) after uniform refinement, with 1024 triangles

Definition 3.5.1 *(Octahedral triangulation of the sphere)*
  The octahedral triangulation of the sphere is defined by the following six vertices:

index	spherical coordinate	index	spherical coordinate
1	$(0, \frac{\pi}{2})$	2	$(0, \frac{-\pi}{2})$
3	$(0, 0)$	4	$(\frac{\pi}{2}, 0)$
5	$(\pi, 0)$	6	$(\frac{3\pi}{2}, 0)$

The corresponding eight elements are defined by

index	vertex index vector	index	vertex index vector
1	$(3, 4, 1)$	2	$(4, 5, 1)$
3	$(5, 6, 1)$	4	$(6, 3, 1)$
5	$(3, 4, 2)$	6	$(4, 5, 2)$
7	$(5, 6, 2)$	8	$(6, 3, 2)$

Remark 3.5.2 *If we consider the unit sphere, with origin $(0,0,0)$, the vertices can be given in cartesian coordinates as follows:*

index	coordinate	index	coordinate
1	$(0, 0, 1)$	2	$(0, 0, -1)$
3	$(1, 0, 0)$	4	$(0, 1, 0)$
5	$(-1, 0, 0)$	6	$(0, -1, 0)$

Another possibility for defining an initial triangulation is the icosahedron. It is a polyhedron with 12 vertices, 20 faces and 30 edges.

Definition 3.5.3 *(Icosahedral triangulation of the sphere)*
  The icosahedral triangulation of the sphere is defined by the following 12 vertices (given in Cartesian coordinates):

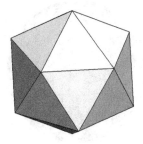 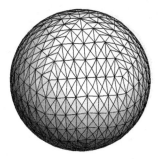

Fig. 3.20. Icosahedral triangulation of sphere: (**left**) initial triangulation, with 20 triangles; (**right**) and after refinement, with 1280 triangles

index	Cartesian coordinate	index	Cartesian coordinate
1	$(0,0,1)$	2	$(\frac{2}{\sqrt{5}},0,\frac{1}{\sqrt{5}})$
3	$(\sqrt{\frac{10-4\sqrt{5}}{25-5\sqrt{5}}},\sqrt{\frac{2}{5-\sqrt{5}}},\frac{1}{\sqrt{5}})$	4	$(-\sqrt{\frac{5+\sqrt{5}}{25-5\sqrt{5}}},\frac{-1+\sqrt{5}}{\sqrt{10-2\sqrt{5}}},\frac{1}{\sqrt{5}})$
5	$(-\sqrt{\frac{5+\sqrt{5}}{25-5\sqrt{5}}},\frac{1-\sqrt{5}}{\sqrt{10-2\sqrt{5}}},\frac{1}{\sqrt{5}})$	6	$(\sqrt{\frac{10-4\sqrt{5}}{25-5\sqrt{5}}},-\sqrt{\frac{2}{5-\sqrt{5}}},\frac{1}{\sqrt{5}})$
7	$(\sqrt{\frac{5+\sqrt{5}}{25-5\sqrt{5}}},\frac{-1+\sqrt{5}}{\sqrt{10-2\sqrt{5}}},-\frac{1}{\sqrt{5}})$	8	$(-\sqrt{\frac{10-4\sqrt{5}}{25-5\sqrt{5}}},\sqrt{\frac{2}{5-\sqrt{5}}},-\frac{1}{\sqrt{5}})$
9	$(-\frac{2}{\sqrt{5}},0,-\frac{1}{\sqrt{5}})$	10	$(-\sqrt{\frac{10-4\sqrt{5}}{25-5\sqrt{5}}},-\sqrt{\frac{2}{5-\sqrt{5}}},-\frac{1}{\sqrt{5}})$
11	$(\sqrt{\frac{5+\sqrt{5}}{25-5\sqrt{5}}},\frac{1-\sqrt{5}}{\sqrt{10-2\sqrt{5}}},-\frac{1}{\sqrt{5}})$	12	$(0,0,-1)$

The corresponding 20 elements are defined by (e.g.)

index	vertex index vector	index	vertex index vector
1	$(3,4,1)$	2	$(4,5,1)$
3	$(5,6,1)$	4	$(6,7,1)$
5	$(7,3,1)$	6	$(3,7,12)$
7	$(11,12,7)$	8	$(7,6,11)$
9	$(10,11,6)$	10	$(6,5,10)$
11	$(9,10,5)$	12	$(5,4,9)$
13	$(8,9,4)$	14	$(4,3,8)$
15	$(12,8,3)$	16	$(8,12,2)$
17	$(9,8,2)$	18	$(10,9,2)$
19	$(11,10,2)$	20	$(12,11,2)$

Remark 3.5.4 *The rather obscure coordinate values in definition 3.5.3 can be circumvented, when considering a sphere with radius $r = \frac{1}{2}\sqrt{5}$. Then, the coordinates can be given in 3D cartesian form by*

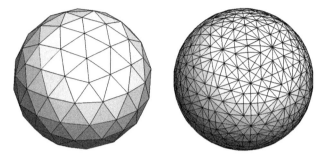

Fig. 3.21. Truncated icosahedral triangulation of sphere (buckyball): (**left**) initial triangulation, with 180 triangles; (**right**) after uniform refinement, with 1440 triangles

index i	x-coordinate	y-coordinate	z-coordinate
1	0.0	0.0	r
2	0.0	0.0	$-r$
3:7	$\cos\left(\frac{2(i-3)\pi}{5}\right)$	$\sin\left(\frac{2(i-3)\pi}{5}\right)$	0.5
8:12	$\cos\left(\frac{\pi+2(i-8)\pi}{5}\right)$	$\sin\left(\frac{\pi+2(i-8)\pi}{5}\right)$	-0.5

Remark 3.5.5 *In spherical coordinates (λ, ϕ, r) the vertex positions can be given numerically and for arbitrary radius r by*

index	spherical coordinate	index	spherical coordinate
1	$(0, \frac{\pi}{2})$	2	$(0, \frac{-\pi}{2})$
3	$(0, 0.4636)$	4	$(\frac{2\pi}{5}, 0.4636)$
5	$(\frac{4\pi}{5}, 0.4636)$	6	$(\frac{-4\pi}{5}, 0.4636)$
7	$(\frac{-2\pi}{5}, 0.4636)$	8	$(\frac{\pi}{5}, -0.4636)$
9	$(\frac{3\pi}{5}, -0.4636)$	10	$(1, -0.4636)$
11	$(\frac{-3\pi}{5}, -0.4636)$	12	$(\frac{-\pi}{5}, -0.4636)$

Note that the icosahedron is the dual to the dodecahedron, a polyhedron with 12 pentagonal faces. Therefore a triangulation of the dodecahedron results in the same grids as a triangulation of the icosahedron. An application using the icosahedral grid to derive a spherical geodesic grid is given in [339].

Finally, another possibility to define the initial triangulation of a sphere is to use the truncated icosahedron (better known as *buckyball* or simply soccer ball). The truncated icosahedron consists of 12 pentagons and 20 hexagons. These polygons are triangulated by adding their midpoint to the vertex list and connecting it to all surrounding polygon vertices. A concise description of the construction of such triangulation is given in [193]. We only give the vertex list for the truncated icosahedron here.

Definition 3.5.6 *(Truncated icosahedral triangulation of the sphere)*

The truncated icosahedral triangulation of the sphere is defined by the following 92 vertices:

index	sph. coord.	index	sph. coord.	index	sph. coord.
1	$(0, \frac{\pi}{2})$	2	$(0.6283, 1.2204)$	3	$(1.8850, 1.2204)$
4	$(-\pi, 1.2204)$	5	$(-1.8850, 1.2204)$	6	$(-0.6283, 1.2204)$
7	$(1.2566, 0.9184)$	8	$(2.5133, 0.9184)$	9	$(-2.5133, 0.9184)$
10	$(-1.2566, 0.9184)$	11	$(0, 0.9184)$	12	$(-0.6283, 0.8141)$
13	$(0.6283, 0.8141)$	14	$(1.8850, 0.8141)$	15	$(-\pi, 0.8141)$
16	$(-1.8850, 0.8141)$	17	$(-0.2376, 0.5409)$	18	$(0.2376, 0.5409)$
19	$(1.0190, 0.5409)$	20	$(1.4942, 0.5409)$	21	$(2.2757, 0.5409)$
22	$(2.7509, 0.5409)$	23	$(-2.7509, 0.5409)$	24	$(-2.2757, 0.5409)$
25	$(-1.4942, 0.5409)$	26	$(-1.0190, 0.5409)$	27	$(0.6283, 0.4636)$
28	$(1.8850, 0.4636)$	29	$(-\pi, 0.4636)$	30	$(-1.8850, 0.4636)$
31	$(-0.6283, 0.4636)$	32	$(1.2566, 0.1887)$	33	$(2.5133, 0.1887)$
34	$(-2.5133, 0.1887)$	35	$(-1.2566, 0.1887)$	36	$(0, 0.1887)$
37	$(0.4220, 0.1725)$	38	$(0.8346, 0.1725)$	39	$(1.6787, 0.1725)$
40	$(2.0912, 0.1725)$	41	$(2.9353, 0.1725)$	42	$(-2.9353, 0.1725)$
43	$(-2.0912, 0.1725)$	44	$(-1.6787, 0.1725)$	45	$(-0.8346, 0.1725)$
46	$(-0.4220, 0.1725)$	47	$(2.7195, -0.1725)$	48	$(2.3070, -0.1725)$
49	$(1.4629, -0.1725)$	50	$(1.0504, -0.1725)$	51	$(0.2063, -0.1725)$
52	$(-0.2063, -0.1725)$	53	$(-1.0504, -0.1725)$	54	$(-1.4629, -0.1725)$
55	$(-2.3070, -0.1725)$	56	$(-2.7195, -0.1725)$	57	$(-\pi, -0.1887)$
58	$(1.8850, -0.1887)$	59	$(0.6283, -0.1887)$	60	$(-0.6283, -0.1887)$
61	$(-1.8850, -0.1887)$	62	$(2.5133, -0.4636)$	63	$(1.2566, -0.4636)$
64	$(0, -0.4636)$	65	$(-1.2566, -0.4636)$	66	$(-2.5133, -0.4636)$
67	$(2.1226, -0.5409)$	68	$(1.6473, -0.5409)$	69	$(0.8659, -0.5409)$
70	$(0.3907, -0.5409)$	71	$(-0.3907, -0.5409)$	72	$(-0.8659, -0.5409)$
73	$(-1.6473, -0.5409)$	74	$(-2.1226, -0.5409)$	75	$(-2.9040, -0.5409)$
76	$(2.9040, -0.5409)$	77	$(1.2566, -0.8141)$	78	$(0, -0.8141)$
79	$(-1.2566, -0.8141)$	80	$(-2.5133, -0.8141)$	81	$(2.5133, -0.8141)$
82	$(1.8850, -0.9184)$	83	$(0.6283, -0.9184)$	84	$(-0.6283, -0.9184)$
85	$(-1.8850, -0.9184)$	86	$(-\pi, -0.9184)$	87	$(2.5133, -1.2204)$
88	$(1.2566, -1.2204)$	89	$(0, -1.2204)$	90	$(-1.2566, -1.2204)$
91	$(-2.5133, -1.2204)$	92	$(0, \frac{-\pi}{2})$		

Remark 3.5.7 *It should be noted that the triangulations obtained from regular triangular refinement (algorithm 3.3.6) from an icosahedron are equivalent to refinement of a truncated icosahedron and to refinement of a dodecahedron. If we start with an icosahedron and refine two levels, and compare the grid with one refinement of a triangulated dodecahedron and finally with a triangulated truncated icosahedron as described in [193], we get the exactly same triangulations. However, with a bisection refinement strategy this is not true and this is the reason for dealing with it here.*

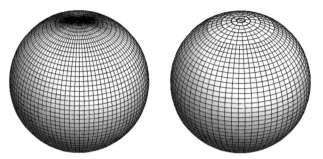

Fig. 3.22. (**Left**) Lat-Lon grid on the sphere. It can be easily seen that the poles form a singularity. (**Right**) The reduced Lat-Lon grid relaxes the problem of small mesh size near the pole

Quadrilateral Initial Triangulations

The simplest and most straight forward quadrilateral triangulation of the sphere is a grid consisting of latitudinal and longitudinal coordinates, a so called *Lat-Lon grid*. Thus a $1^{\deg}$-grid consists of 360×180 vertices. However, the two poles $(0, 90)$ and $(0, -90)$ are singular points in this grid. In fact, all grid cells adjacent to the north and south poles are triangles, since all points $(\lambda, \pm90)$ with $\lambda \in \{0, 360\}$ coincide. Furthermore, many numerical algorithms are restricted in their stability by the mesh size of the smallest grid cell in the mesh. The mesh size at the poles in the Lat-Lon grid approaches 0, which poses severe problems to such algorithms. A $4^{\deg} \times 4^{\deg}$ Lat-Lon grid is depicted in fig. 3.22.

One possibility to circumvent the small mesh size near the poles is the *Reduced Lat-Lon grid*, an example of which is also depicted in fig. 3.22. Here the mesh size is doubled (several times) in the high latitudes, yielding an almost uniform mesh size on the whole sphere. Note that the pole singularity is not removed by this remedy. A semi-Lagrangian solution technique for the shallow water equations on a reduced grid can be found in [259].

A combination of spherical grid and spherically projected unit square grid at the poles, a so called *capped sphere* has been proposed by Lanser et al. [252], based on ideas dating back to Phillips, Starius and Browning et al. [74, 317, 318, 319, 375]. This grid avoids the diminishing mesh size at the poles by introducing a regular grid at the poles and combining this grid with the Lat-Lon grid in mid latitudes in an admissible way. Following the notation in [252], a capped sphere is defined by a Lat-Lon grid in the spherical coordinate range of $0 < \lambda < 2\pi$ and $\phi_{\min} < \phi < \phi_{\max}$, where $0 > \phi_{\min} > -\frac{\pi}{2}$ and $\phi_{\max} = -\phi_{\min}$. When we denote N_{λ} the number of grid points in zonal direction (i.e. the number of longitudinal lines) and N_x the number of grid points in x-direction of the cap grid (we assume $N_x = N_y$, i.e. the number of grid points in x and y-direction to be identical) then in order to obtain an admissible mesh, we need to require $N_{\lambda} - 1 = (N_x - 1) * 4$. Finally, the cap grid is

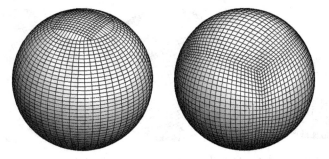

Fig. 3.23. (**Left**) The capped sphere combines two regular unit square grids projected to the poles with a Lat-Lon grid. (**Right**) The cubed sphere is a projection of a rectangular grid on the surface of a cube to the sphere. It exposes eight exceptional points, four of them visible in this figure

given in Cartesian coordinates and projected to the (unit) sphere. For the example given in fig. 3.23 we use $N_x = 15$, $N_\lambda = 57$, $\phi_{\min} = -0.35\pi$ and $(x, y, z) \in [-0.4, 0.4] \times [-0.4, 0.4] \times \{-1, 1\}$.

The *cubed sphere* is another quadrilateral mesh that avoids the pole problem [345]. It is relatively easy to construct: We project a cube composed of six quadrilateral faces to the sphere, for example by a spherical projection. We start with the grid coordinates given in cartesian coordinates. For an equidistant grid this is given by

$$
\begin{aligned}
x_i^{[n]} &= -1 + i \cdot \Delta x, & y_i^{[n]} &= -1, & z_i^{[n]} &= -1 + i \cdot \Delta x, \\
x_i^{[f]} &= -1 + i \cdot \Delta x, & y_i^{[f]} &= 1, & z_i^{[f]} &= -1 + i \cdot \Delta x, \\
x_i^{[l]} &= -1, & y_i^{[l]} &= -1 + i \cdot \Delta x, & z_i^{[l]} &= -1 + i \cdot \Delta x, \\
x_i^{[r]} &= 1, & y_i^{[r]} &= -1 + i \cdot \Delta x, & z_i^{[r]} &= -1 + i \cdot \Delta x, \\
x_i^{[b]} &= -1 + i \cdot \Delta x, & y_i^{[b]} &= -1 + i \cdot \Delta x, & z_i^{[b]} &= -1, \\
x_i^{[t]} &= -1 + i \cdot \Delta x, & y_i^{[t]} &= -1 + i \cdot \Delta x, & z_i^{[t]} &= 1.
\end{aligned}
$$

Here, $\Delta x = \frac{2}{N}$, N is the number of grid points in each direction, $i = 0 : N - 1$ is the grid point index. Superscripts $(\cdot)^{[n,f,l,r,b,t]}$ denote near, far, right, left, bottom, and top faces. A basic cubed sphere mesh is shown in fig. 3.23.

It should be noted that there are diverse projection methods for the cubed sphere that result in different uniformity properties of the grid cells as pointed out in [297]. We are not going to dive too deeply into the details here but refer to the literature. See [387] for a spectral element method on the cubed sphere.

The *Yin-Yang grid* represents an overset gridding method [228]. That means two different grids are composed to construct a mesh covering the whole sphere. This technique is often termed composite mesh generation [84]. In case of the Yin-Yang grid, the complete mesh is constructed of two rotated

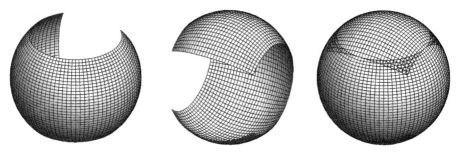

Fig. 3.24. (**Left and center**) The Yin Yang grid consists of two separate grids, composed to one overlapping mesh (**right**), thereby omitting the pole singularity in mesh size. However it adds the problem of consistency in the overlap region

partial Lat-Lon grids, the Yin and the Yang grids resp. The Yang grid can be obtained by rotating the grid points of the Yin grid with the following matrix operation

$$\mathbf{x}_{\text{Yang}} = R \cdot \mathbf{x}_{\text{Yin}}$$

where $\mathbf{x}$. is the cartesian coordinate of a grid point in either the Yin or the Yang grid and R is the rotation matrix, given by

$$R = \begin{bmatrix} -1 & 0 & 0 \\ 0 & 0 & 1 \\ 0 & 1 & 0 \end{bmatrix}.$$

We are left with defining the Yin grid. This is just the spherical $j_{\max} \times k_{\max}$-grid given by coordinates (λ_j, ϕ_k), $j = 0 : j_{\max} - 1$, $k = 0 : k_{\max} - 1$ and

$$\lambda_j = \lambda_{\min} + j \cdot \Delta\lambda,$$
$$\phi_j = \phi_{\min} + j \cdot \Delta\phi,$$

where $\Delta\lambda = (\lambda_{\max} - \lambda_{\min})/(j_{\max} - 1)$, $\Delta\phi = (\phi_{\max} - \phi_{\min})/(k_{\max} - 1)$, $\lambda_{\min} = -\frac{\pi}{4} - \omega$, $\lambda_{\max} = \frac{\pi}{4} + \omega$, $\phi_{\min} = -\frac{3\pi}{4} - \omega$, $\phi_{\max} = \frac{\pi}{4} + \omega$. ω denotes the overlap of the two meshes. Figure 3.24 shows the Yin and the Yang grid together with the composite grid for $\omega = 0$. There are several modifications, for example for non-overlapping matching Yin and Yang grids or for different methods to minimize the overlap.

4

Data Structures for Computational Efficiency

Each adaptive algorithm for solving atmospheric flow problems can be separated into two basic parts:

1. grid generation and management,
2. numerical computations discretizing the differential operators.

While the first part consists of mainly integer operations and database management, the second represents the number crunching part. So, the grid generation part needs tree data structures or database-like data structures (hash tables or heaps) for efficient creation, destruction, and search of mesh items. Conversely, in the numerical part, one tries to achieve consecutive data structures which are well blocked such that hierarchical memory architectures can be utilized beneficially. Many AMR methods mix both parts in that they try to refine the mesh in blocks that are large enough to do efficient computations. Additionally the blocks are self similar, such that operators, defined on a block can be re-used on every level of the grid. These operators can be optimized for performance on corresponding processor architectures.

Another class of applications strictly separates the above two basic parts by introducing a gather and scatter step between the grid generation part and the numerical part. This principle is depicted in fig. 4.1.

4.1 Data Structures for Grid Management

In order to assess data structures for grid management, let us first recall an abstract view of a grid. In adaptive atmospheric computations, grids consist of *nodes*, *edges*, *faces*, and *cells*. Most computations take place on nodes and cells, however, in Finite Volume techniques for example, edges/faces between cells play a crucial role as well.

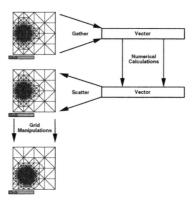

Fig. 4.1. Two phases of an adaptive algorithm, clearly separated by a gather and scatter between numerical data structures and grid data structures

Definition 4.1.1 *(Grid Atoms)*

Node: A node v_i is a point in the computational domain, defined by its coordinates $\mathbf{x}_i \in \mathcal{G} \subset \mathbb{R}^d$, d the space dimension ($d = 1, 2$, or 3).

Edge: An edge e_j is a (possibly curved) line connecting two nodes v_1^j and v_2^j, defined by the pair of nodes (v_1^j, v_2^j).

 Note that in a 2D grid, edges are the interfaces between each two grid cells.

Face: A face τ_k is a polygon in $\mathbb{R}^2$, defined either by the nodes v_i^k, $i = 1 : K$ or the edges e_j^k, $j = 1 : K$.

 Note that in a 2D grid, faces represent the grid cells. Note further, that in a 3D grid, faces are the interfaces between each two cells.

Cell: A cell τ_l^{3D} is a polyhedron in $\mathbb{R}^3$, defined by either its vertices (nodes) v_i^l, $i = 1 : L_v$, or by its edges e_j^l, $j = 1 : L_e$, or by its faces τ_k^l, $k = 1 : L_\tau$.

 It is more common to take either the nodes, or the faces as defining items for cells.

Remark 4.1.2

1. For grids in 1D edges represent the cells, while nodes are the interfaces between each 2 cells. In 2D faces represent the cells, while in 3D the definition of cells is consistent with the definition 4.1.1.
2. The number of adjacent grid items also depends on the dimension. For $d = 1$, nodes can have 2 adjacent edges, while for $d = 2$, there can be more. Similarly, an edge can have two adjacent faces for $d = 2$, while for $d = 3$ there might be many more.
3. We do not consider grids for $d > 3$ here.
4. While this abstract list of grid items is valid for all kinds of grids, including quadrilateral locally structured ones, it is probably not as useful for structured as for unstructured grids.

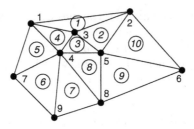

Fig. 4.2. Example grid with numbered nodes (*regular typeface*) and cells (*italic typeface, encircled*)

While the grid atoms defined just above are the building blocks of a grid, the definition of *connectivity* makes them a grid. There are in principle two ways to define the connectivity: One *list-oriented* and one *object-oriented* approach. The list-oriented approach originates from times where most technical software was implemented in FORTRAN77 where no dynamic data allocation was available and data structures were essentially restricted to arrays.

Definition 4.1.3 *(List-oriented grid)*
   A list-oriented grid data structure can be given by three lists:

1. **Coordinates:** *a list with coordinates, this is a $(d \times N)$-array, where d is the number of dimensions and N the number of nodes.*
2. **Cells:** *a list of indices to the cell's nodes (of size $K \times M$, where M is the number of cells, K the number of nodes per cell).*
3. **Connectivity:** *a list of indices to the cell's neighbors of size $L \times M$, where L is the number of faces.*

Example 4.1.4 *An example grid is given in fig. 4.2. The corresponding cell and connectivity arrays according to definition 4.1.3 are given in the following equation. $N = 9$, $M = 10$, $d = 2$, and $K = L = 3$. We denote the cell array by* **el** *and the connectivity array by* **ne**. *If the cell is at the boundary of the domain and has no neighbor at one side, we indicate this by an entry 0 in the connectivity array.*

$$
\mathbf{el} = \begin{bmatrix} 1&3&4&4&7&7&9&8&8&5 \\ 3&5&5&3&4&9&8&5&6&6 \\ 2&2&3&1&1&4&4&4&5&2 \end{bmatrix},
$$

$$
\mathbf{ne} = \begin{bmatrix} 2&10&2&1&4&7&8&3&10&0 \\ 0&1&4&5&0&5&6&7&8&2 \\ 4&3&8&3&6&0&0&9&0&9 \end{bmatrix}.
$$

Note that the three lists of definition 4.1.3 define a grid completely. One can easily see, that this type of data structure needs a lot of care when introducing adaptive refinement with changing numbers of cells. Additionally, when removing cells due to coarsening the mesh, gaps will occur in the arrays raising the need for packing/defragmentation.

A more convenient data structure is an object oriented one. There is much more choice on the design here. Therefore, we will restrict ourselves to one concept, used in the grid generation software amatos [38].

Definition 4.1.5 *(Object-oriented grid)*
An object oriented grid data structure consists of grid atoms according to definition 4.1.1, where the interface atoms (edges in 2D and faces in 3D) have information about their two neighbors.

Example 4.1.6 *Looking again at fig. 4.2, we have*

- $N = 9$ *node objects defined by their coordinates (not given here);*
- $L_e = 18$ *edge objects defined by their nodes and knowing their neighbors:*

item	nodes		neig's		item	nodes		neig's		item	nodes		neig's	
1	1	2	1	0	2	1	3	4	1	3	3	2	2	1
4	7	1	5	0	5	1	4	5	4	6	4	3	3	4
7	3	5	3	2	8	5	2	10	2	9	2	6	10	0
10	7	4	6	5	11	4	5	8	3	12	5	6	9	10
13	7	9	0	6	14	9	4	7	6	15	4	8	7	8
16	8	5	9	8	17	9	8	0	7	18	8	6	0	9

- $M = 10$ *elements, defined by their edges:*

item	1	2	3	4	5	6	7	8	9	10
edge1	2	7	5	6	10	13	17	16	18	12
edge2	3	8	7	2	5	14	15	11	12	9
edge3	1	3	6	5	4	10	14	15	16	8

While the object-oriented grid data structure looks less compact (and in fact uses slightly more memory) it is much more versatile. All atoms can be stored in consecutive lists, linked lists, or hash tables, and removing an atom due to coarsening of the grid, leads to only local updates. In fact, only the edges of the concerned elements have to be updated. No defragmentation is necessary and the information needed for the update is immediately at hand. Objects can be combined in derived data structures thus, a convenient data movement in multi-process environments is possible. For more details on grid data structures see chap. 14 of [391]

Finally, we need to create, find and delete elements in an efficient way. From the description of refinement strategies in chap. 3 it is obvious that we will use tree-like data structures for this task.

Definition 4.1.7 *(Tree)*
A tree is either

- *empty (no nodes), or*
- *a root and has zero or more sub-trees.*

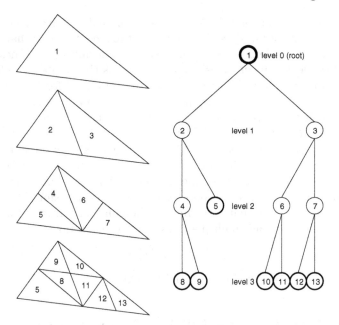

Fig. 4.3. An example of a refinement tree, induced by triangular bisection

A root *is a node in the tree, that has no parents. While a* node *is either a* leaf *that has no children or an* internal node *with parent and children.*

Setting up a tree data structure for grid refinement, we need to define a root and then sub-trees. Each element of the initial (coarse) triangulation can be defined a root of a tree. Then, each child of the root element is root of a sub-tree.

Remark 4.1.8 *(Binary tree – bisection refinement)*
The bisection refinement strategy (algorithm 3.4.3 or 3.3.1) yields a binary tree, *that is a tree where each node has at most two children. In fact, if a node (triangle) has children, then (with bisection) it has exactly two children.*

An example of a binary tree together with the corresponding mesh is given in fig. 4.3.

Remark 4.1.9 *(Quad tree – regular refinement)*
The 2D regular triangular refinement (algorithm 3.3.6) and the 2D regular quadrilateral refinement (algorithm 3.3.11) both yield a quad tree, *where each node has at most four children.*

Remark 4.1.10 *(Oct tree – 3D hexahedral refinement)*
The 3D quadrilateral refinement (algorithm 3.4.5) yields an oct tree, *where each node has at most eight children.*

Now, when searching one element of the triangulation, given a coordinate, a fast and efficient way to do so is to traverse the tree. Suppose that we start with only one initial element (root) and have a uniformly refined mesh. Then searching for a leaf element takes l steps, where l is the number of refinement levels. The total number of elements (neglecting closures) for l refinements is $M = b^l$, where we have $l = 0$ for the root element, and $b = 2, 4, 8$ the basis according to the correspondences in the above remarks. Thus, a tree search takes $l = \log_b(M)$ steps, and is almost optimal.

Refinement of elements is possible only at the leafs of the tree data structure. When an element is refined the tree is expanded. Coarsening means cutting of sub trees. A locally refined mesh can therefore have different depths of a tree. Thus, the above calculation of the computational effort for searching is an upper bound. Note that if the initial triangulation consists of more than one element, then we have multiple trees, thus a search takes

$$M_0 \cdot \log_b(M)$$

steps, where M_0 is the number of initial triangles.

4.2 Data Structures for Efficient Numerical Calculations

Data structures for numerical calculations are likely to be well suited for optimization if they are in vector like shape. There is a lot of experience with vectorization. Most vectorized code structures can also easily be parallelized. However, this is not the whole story. This section has to investigate efficient neighborhood preserving data structures for adaptive grid refinement.

While the description of grid data structures in the previous section was valid for both unstructured triangular as well as semi-structured quadrilateral refinement strategies, we have to consider structured and unstructured grids separately here. For the semi-structured meshes, a common technique of optimization is the definition of operators on grid patches. We will deal with this in the next section. Here, we will focus on simple optimizations when working on unstructured triangular meshes.

4.2.1 Vectorizable Data Structures

There are several abstract tasks that have to be executed on grid unknowns. Let us denote by $\rho(\mathbf{x}_i, t_j)$, $i = 1 : N, j = 0 : I$, a value ρ at unknown position $\mathbf{x}_i$ and time step t_j.

1. update unknowns with independent function values: $\rho(\mathbf{x}_i, t_j) = f(\mathbf{x}_i, t_j)$,
2. update unknowns with spatially dependent function values:
   $\rho(\mathbf{x}_i, t_j) = F(\mathbf{x}_i, t_j; \rho(\chi, t_j))$, where $\chi \subset \{\mathbf{x}_i | i = 1 : N\}$.
3. update unknowns with space-time dependent function values:
   $\rho(\mathbf{x}_i, t_j) = F(\mathbf{x}_i, t_j; \rho(\chi, \tau))$, where $\chi \subset \{\mathbf{x}_i | i = 1 : N\}$ and $\tau \subset \{t_j | j = 1 : I\}$.

Example 4.2.1 *(Finite difference operators)*
*   We can easily formulate a backward time difference ∂_t or a first order spatial centered difference operator ∂_x by the above abstract notation:*

$$\partial_t \rho(\mathbf{x}_i, t_j) = \frac{\rho(\mathbf{x}_i, t_j) - \rho(\mathbf{x}_i, t_{j-1})}{\Delta t} =: F_t\left(\mathbf{x}_i, t_j; \{\rho(\mathbf{x}_i, t_j), \rho(\mathbf{x}_i, t_{j-1})\}\right),$$

$$\partial_x \rho(\mathbf{x}_i, t_j) = \frac{\rho(\mathbf{x}_{i+1}, t_j) - \rho(\mathbf{x}_{i-1}, t_j)}{2\Delta x} =: F_x\left(\mathbf{x}_i, t_j; \{\rho(\mathbf{x}_{i+1}, t_j), \rho(\mathbf{x}_{i-1}, t_j)\}\right).$$

If $\mathbf{x} = (\mathbf{x}_1, \ldots, \mathbf{x}_N)$ is the consecutive vector of unknown positions and $\mathbf{r}_j = (\rho_1, \ldots, \rho_N)$, where $\rho_i = \rho(\mathbf{x}_i, t_j)$ is the consecutive vector of values, than the first task can be efficiently vectorized (and therefore parallelized) by the vector operation

$$\mathbf{r}_j = F_t(\mathbf{x}, t_j).$$

In order to formulate a vectorizable version of the second task above, we need to divide χ into $\chi_o = \{\mathbf{x}_l | l < i\}$ and $\chi_n = \{\mathbf{x}_l | l \geq i\}$. Provided that F is a linear operator, we can write (with the above vector notations)

$$\mathbf{r}_j = F(\mathbf{x}, t_j; \rho(\chi_o, t_j)) + F(\mathbf{x}, t_j; \rho(\chi_n, t_j)).$$

When computing this on a parallel or vector computer, a data dependency can occur for the first term on the right hand side. However, this type of data dependency can be treated by common techniques [205].
    For the third task in the list above, a similar separation ansatz can be chosen:

$$\mathbf{r}_j = F(\mathbf{x}, t_j; \rho(\chi_o, t_j)) + F(\mathbf{x}, t_j; \rho(\chi_n, t_j)) + \sum_{t_k \in \tau} F(\mathbf{x}_i, t_j; \rho(\chi, t_k)).$$

A data dependency can again be observed in the first term and has to be resolved.
    With the formalization and the examples, most numerical operations in atmospheric modeling software can be covered. And with a vectorizing programming model, we can optimize these operations efficiently.

4.2.2 Maintaining Data Locality – Reordering Schemes

Another aspect of optimization is data locality. So, in addition to vector-like data structures, data in the vectors should preserve some kind of locality properties when mapped from a d-dimensional mesh to a 1-dimensional vector (see for example [284]).

Example 4.2.2 *(Data locality for finite difference Laplace operator)*
*   The second order Laplace operator is defined by $\Delta\rho = \frac{\partial^2 \rho}{\partial x^2} + \frac{\partial^2 \rho}{\partial y^2}$. It can be discretized by using forward and backward finite difference operators with the well known five point stencil (we omit the time index, and use the 2D position $\mathbf{x}_i = (x_i, y_i)$):*

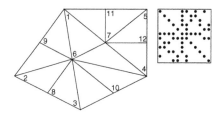

Fig. 4.4. A simple triangular mesh, created by a bisection refinement. The refinement induces a numbering of nodes as given in the left mesh picture. The corresponding connectivity matrix (**right**) has a non-local structure

$$\Delta\rho(x_i, y_i) \approx \frac{1}{\Delta x^2} \left[4\rho(x_i, y_i) - \rho(x_{i+1}, y_i) - \rho(x_{i-1}, y_i) \right.$$
$$\left. - \rho(x_i, y_{i+1}) - \rho(x_i, y_{i-1}) \right].$$

Applying this operator to a vector $r_j = (\rho_1, \ldots, \rho_N)$ results in a corresponding matrix A with entries $\frac{4}{\Delta x^2}$ in the diagonal and $\frac{-1}{\Delta x^2}$ in four sub-diagonals. Preserving locality means keeping the sub-diagonals close to the diagonal.

The example suggests a common technique for neighborhood preservation in mapping d-dimensional data to 1-dimensional vectors, i.e. matrix reordering. Well-known techniques for data reordering are the Cuthill-McKee algorithm [104], and the minimum degree ordering [161]. We illustrate the reordering strategies by a simple grid and connectivity matrix[1] example illustrated in fig. 4.4.

The Cuthill-McKee algorithm orders a graph by the following procedure. Note that we use the one-to-one correspondence of a graph to the grid, where each edge of our grid corresponds to an edge in the connectivity graph, referred to in the algorithm.

Algorithm 4.2.3 *(Cuthill-McKee reordering)*
Let n_i, $i = 1 : N$ be nodes of a graph and let e_j, $j = 1 : M$ be the edges connecting nodes. A node n_l is called adjacent *to node n_i, if there exists an edge e_k connecting n_i and n_l.*

1. *Start the procedure by selecting a starting node (e.g. the one with lowest degree). Relabel it with 1 (obtaining n_1).*
2. *Relabel all nodes adjacent to n_1 in consecutive order (taking the nodes with lower degree first). These nodes are called level 1 nodes, since they have distance 1 from n_1 in terms of graph connections.*
3. *Repeat step 2 for each level k node, obtaining level $k+1$ nodes, $k = 1, 2, \ldots$.*

Note that algorithm 4.2.3 terminates when all connections have been traveled which means that all nodes have been numbered (since our grid is equivalent to a connected graph). Cuthill and McKee could show that this algorithm results

---

[1] For a formal definition of a connectivity matrix, see sect. 5.1.

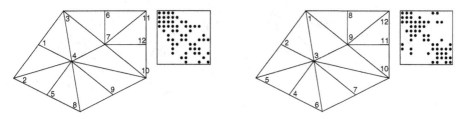

Fig. 4.5. (Left) The Cuthill-McKee reordering yields low bandwidth (high locality) connectivity; **(right)** the minimum degree reordering looks less favorable in this example

in a sparse matrix structure with almost minimum bandwidth. In other words, nodes that are adjacent are close to each other in the consecutive 1D ordering as well. The effect of reordering is depicted in fig. 4.5.

A second and also well introduced reordering scheme is called (approximate) minimum degree ordering. It was introduced by Rosen [346] for minimizing the bandwidth of sparse matrices and is still used in modern software (see e.g. [109]). The original algorithm reads

Algorithm 4.2.4 *(Minimum degree reordering)*

1. *Determine those nodes that cause the largest bandwidth in the sparse matrix (i.e. that are farthest away from each other in the consecutive numbering scheme).*
2. *Determine, whether an interchange of the higher node index, say i, with some other node index, say j, yields a smaller bandwidth (i.e. moves it closer to its neighbors).*
3. *If yes, interchange i and j.*
4. *Repeat steps 1. to 3. until no further interchanges are possible.*

There are much more efficient ways to compute the minimum degree reordering (see e.g. [110]); however, this discussion is not in the scope of our presentation. We used an implementation of the minimum degree column ordering as given in MATLAB [8] to compute the reordering given in fig. 4.5.

A very efficient and in most cases even better reordering mechanism stems from space-filling curves (SFC). A geometric construction of SFC was originally proposed by Hilbert [203], based on ideas by Peano [310], for proving a set theoretical problem. A space-filling curve passes through all points of a multi-dimensional domain, thereby, mapping the multi-dimensional index space to a one-dimensional set (curve). One can show that SFC preserve neighborhood relations. Moreover, space-filling curves are well suited for parallelization, since they induce a partition of the domain (see sect. 5.1).

It is easy to calculate SFC for adaptive grids. In fact, for triangular bisection, the index calculation consists of just one bit-manipulation per triangle [43]. The SFC for triangular bisection grids corresponds to a Sierpinsky curve

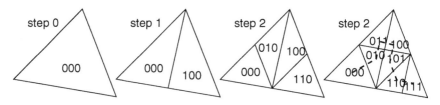

Fig. 4.6. Series of steps in the construction of a space-filling curve in a locally bisection refined triangular mesh

[45, 429]. In contrast, a well suited SFC for quadrilateral grids is the Hilbert curve, used for example by Roberts et al. [344] and by Griebel and Zumbush [178]. Hilbert's curve can also be constructed by recursion [73]. Several alternative patterns are proposed in [270].

The following data have to be known a priory:

a) the number of triangles in the initial triangulation N_0,
b) the maximum number of refinement levels l.

With these data, for each element we need a bit structure of length $b = \log_2(N_0) + l$. While the first $b - l$ bits are used for consecutively numbering the initial elements arbitrarily, each level is then represented by an additional bit. To illustrate the following algorithm, observe fig. 4.6.

Algorithm 4.2.5 *(Space-filling curve for 2D bisection)*
Let τ^k be a cell on level k of the grid, and we denote with τ_i^k, $(i = 1 : 2)$ both daughters of cell τ^{k-1}. For simplicity, we assume only one cell τ^0 in the initial triangulation, therefore $b = l$.

1. *The algorithm starts with a zero bitmap of length b in τ^0.*
2. FOR *each level $(k = 1 : l)$* DO:
    a) *copy the mother's (τ^{k-1}) bitmap to both daughter cells (τ_i^k);*
    b) *determine left or right side cell τ_e^k according to the level:*

$$\begin{cases} \tau_e^k = \text{left, if} & \mod(k, 2) = 0, \\ \tau_e^k = \text{right, if} & \mod(k, 2) = 1; \end{cases}$$

    c) *set the k-th bit of daughter τ_e^k to 1.*
3. END FOR

The effect of the SFC reordering on data locality, i.e. the connectivity matrix is shown in fig. 4.7. Although the SFC reordering causes some entries far from the diagonal, which means far away in memory, in most cases the locality is very well maintained. Note that for many numerical computations involving sparse matrices the inverse of the SFC numbering is even better, since it prevents creation of fill-in.

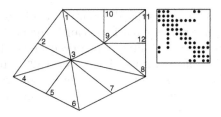

Fig. 4.7. SFC reordering yields good data locality as can be seen from the connectivity matrix

Vector-like data structures with neighborhood-preserving ordering are a handy tool for optimized numerical calculations. One technique to gain performance in adaptive (unstructured) programs is therefore to gather data from grid data structures into vector-like data structures before entering the numerical computation parts of the program. This paradigm of code structuring has been used in the grid generation software `amatos` described in sect. 4.5 below.

4.3 Working With Grid Patches

In contrast to triangular refinement strategies, where cells are created individually, in block-structured quadrilateral methods, a whole block of cells is the building block of the mesh. When using self-similar blocks (i.e. the same number of cells in each block, regardless of the refinement level) discrete mathematical operators like difference operators (for finite difference methods) or flux operators (for finite volume methods) can be defined in an abstract and optimized way.

Definition 4.3.1 *(Grid Patch)*
A grid patch is an abstract data structure defined by the following properties:

1. *the **position**, for example the (x, y, z)-coordinate of the lower left corner;*
2. *the grid **level** within the refinement hierarchy (alternatively or additionally the grid spacing can be used as defining parameter);*
3. *the number of **grid-points** in each space dimension (in other words, the size of the grid patch).*

There may be additional information like neighborhood relations, ghost cells, mother/daughter grid patches, etc. attached to the data object grid patch.

With this definition an AMR supporting software can define iterators that handle all grid patches in an adaptively refined mesh. By defining grid patches by some (few) characteristic parameters, a simplified handling of grids is possible.

Now, mathematical operators like differential operators that are needed to discretize the underlying model equations, can be defined patch wise. Using grid patches of a limited variability in size even allows for hardware optimized implementations of operators. Jablonowski [221] reports on optimal patch sizes, where the cache size of the microprocessor architecture plays an important role. Balancing the size of a grid patch is an optimization problem stated in the following way:

Remark 4.3.2 *(Balancing grid patch size)*
The minimum time required for solving a given problem by an AMR method with grid patches is obtained by choosing the optimum between the following competing functions:

1. *the time per unknown decreases with increasing patch size up to a saturation limit (usually the cache size);*
2. *the overall time for solution grows linearly with the number of unknowns;*
3. *the number of (overhead) unknowns grows with increasing patch size;*
4. *the number of patches grows with the number of refinement regions/ localized features.*

Note that this type of balancing is not necessary (and not applicable) for triangular grids with refinement strategies described in chap. 3. On the other hand, optimized algorithmic blocks defined on grid patches are not possible on triangular grids either. Giving up some flexibility in the grid adaptation yields enhanced simplicity and performance in grid operations.

4.4 Software Packages for Adaptive Grid Management

Several authors have contributed packages that ease the implementation of adaptive models. Many of these tools are either purely research tools or have been implemented for a broad application spectrum. One such tool is called *ALBERTA* and maintained by Kunibert Siebert and Alfred Schmid [354, 353]. It is a finite element toolbox that is well structured in a hierarchical way and is well suited for academic research and education. However, since it is not parallelized, it cannot be used for real life computationally intense applications. It can be regarded as a reference for data structures and code design.

Several other tools for triangular and quadrilateral grid generation and finite element support are available. We want to reference only a few. A comprehensive overview of finite element software can be found at the *Internet Finite Element Resources* [424]. *QMG* is a research tool that works within either MATLAB or with a Tcl/Tk interpreter. It aims at solving finite element discretized problems with complex geometries in 2D and 3D [399]. *CUBIT* from Sandia National Laboratory is a more production oriented tool for finite element simulation [306]. *CUBIT* is a collection of tools for generating unstructured 2D and 3D meshes, based on tetrahedra or hexahedra, and geometry preparation. It is mainly built to generate meshes from given geometries. A comprehensive collection on mesh generation tools can be found online [307, 355].

AMROC by Ralf Deiterding is a block structured AMR tool implemented in an object oriented manner [114, 113]. It is parallelized for distributed memory architectures and serves as an object oriented framework for flow solvers. An example is given for LeVeque's *CLAWPACK* (see below). *SAMRAI* is a framework for structured adaptive mesh refinement, developed at Lawrence Livermore National Lab [416].

Collela et al. at Lawrence Berkeley National Laboratory have developed *CHOMBO* a quite sophisticated tool for parallel structured adaptive mesh refinement [98, 97]. *Cart3D* is a Cartesian mesh based grid generation and computational fluid dynamics package [4]. Of course, the mother of all block structured AMR packages *CLAWPACK* (formerly *AMRCLAW*) by LeVeque and coworkers is available free of charge for research purposes [262]. Blikberg parallelized an older version of AMRCLAW in an application to solve the shallow water equations [63]. The latest version (4.2 at the date of writing this) is parallelized by the message passing interface (MPI) and is therefore capable of computing on distributed memory machines.

There are hardly any grid generation packages available that generically support two-dimensional spherical geometries as required by global or large scale atmospheric modeling. For triangular unstructured meshes such a tool, amatos, is presented in the next section [38, 44, 32]. A tool that supports spherical geometries with block structured meshes is used by Jablonowski in her adaptive atmospheric model [221]. The blocked spherical adaptive mesh refinement tool has been developed by Oehmke and Stout [304].

4.5 Example for a Grid Handling Software Package – amatos

amatos is designed to reduce the code complexity for the application programmer. At the same time, amatos aims at efficiency, parallelization and ease of use. Applications that have been realized using amatos include stationary problems of simple potential calculations, tracer advection in 2D and 3D, and implementation of dynamical cores for weather prediction and climate simulation.

amatos supports different numerical schemes but has been primarily designed with semi-Lagrangian respectively Lagrange-Galerkin schemes in mind. The library supports finite element (FEM) type spatial discretizations, including spectral elements.

This section introduces the key techniques behind the grid generating library. These topics cover grid generation, efficient data structures, including space-filling curve (SFC) ordering, interpolation of scattered data, including radial basis function (RBF) interpolation, FEM support issues, and computational geometry issues. For a detailed technical documentation of `amatos` the reader is referred to [32, 37].

Main Data Structures

We start this technical description with an introduction of the main data structures of `amatos`. While implemented in Fortran 90, the design of `amatos` tried to adhere to object oriented principles. The objects of a mesh are *nodes*, *edges*, *elements* (or faces), and *tetrahedra*. Each of these objects is defined by a certain set of data, has attributes and may be linked to other items. Thus, each grid item is represented by a derived structured data type with substructures `def`, `att`, and `lnk`, for definition, attributes, and links resp.

The node objects are defined by their physical coordinates `r_coor`. Additionally each node has a unique identifying index `i_indx`. Attributes of a node are it's time stamp, an indicator for it's time of occurrence, it's patch (surrounding elements), the edge index it divides, and it's physical values. Finally a node links to it's periodic partner, when periodic boundary conditions are applied.

Edge objects are defined by their node indices `p_node` and again by a unique identifier `i_indx`. Attributes of an edge are among others the time stamp, it's state of refinement, it's boundary condition (if it is a boundary edge), it's values, and if parallelized it's processor number. Links are set to daughter edges (forming a 1-way binary tree) and periodic partner.

Elements (faces) are defined by their nodes (vertices) `p_node` or alternatively by their edges `p_edge`. `amatos` uses both definitions redundantly (and takes care for consistency). Element's attributes are time stamp, indicator for it's time of occurrence, state of refinement, level of refinement, it's SFC index, and it's values. Links are set to the mother and daughter elements (forming a 2-way binary tree structure).

In principle, hash tables or linked lists would have done the job of managing all items elegantly, but for the sake of quick random access, pointer arrays have been chosen to store the addresses of all items. The main disadvantage of this management strategy is the need for packing/defragmenting these arrays from time to time, when coarsening has been conducted.

Several index lists are maintained for convenience and to achieve higher efficiency for certain grid access structures. Index lists formally are just permutation or selection vectors. There is an index list for all elements of the finest local level (i.e. all elements that form the actual computational mesh), there is a list for all boundary edges, etc.

Mesh Refinement Strategy

Currently, there are triangular (for 2D) and tetrahedral meshes implemented in the `amatos`-framework. In order to refine a given triangular/tetrahedral mesh cell, amatos employs a bisection refinement strategy, originally proposed by Rivara [341] and refined by Bänsch [25]. This strategy is based on bisecting a marked edge and a marking algorithm that prevents small angles in the refined mesh, as described in sect. 3.3. With this algorithm no special treatment is necessary for hanging nodes, since hanging nodes can be avoided by an iterative closure. A space-filling curve for ordering the mesh cells can be constructed extremely easily as shown in [43] and described in sect. 4.2.2. With uniform refinement every two levels all edges are halved. One disadvantage of this refinement strategy is the potential for criss-cross meshes that have shown numerical instabilities in certain situations (see for example [300]). For three-dimensional tetrahedral grids, again an extension to the 2D bisection algorithm proposed by Bänsch is followed. With these refinement and coarsening strategies, amatos is capable of creating arbitrarily refined meshes.

Gather/Scatter and Efficient Data Structures

amatos clearly separates the two phases of an adaptive calculation:

1. grid generation/modification phase,
2. numerical calculation phase.

Calculations in the grid generation phase are performed on trees of grid atoms (i.e. tetrahedra, triangles, edges, nodes). Each atom in the grid has a unique identification index and can be accessed efficiently.

Once the mesh has been computed, a gather step collects data required for the numerical computation from the grid data structures within amatos. The result of the gather is a vectorized data structure, e.g. a vector of all unknowns (with dimension N, where N is the total number of unknowns), or a matrix of all coordinates (with dimension (d,N), where d is the coordinate dimension and N as above). Once the numerical computing phase has been completed, a scatter step has to be performed to write back new data into mesh data structures. Finally, the mesh can be manipulated, and data is automatically propagated with the mesh. Figure 4.1 at the beginning of this chapter illustrates this principle.

While the gather/scatter steps add a certain overhead to the whole procedure, this is compensated by the higher efficiency in both the grid generation phase and computational phase. In many of our practical problems, we see less than 1% overhead introduced by gather/scatter.

It should be mentioned that the gather/scatter routines contain a lot of intelligence, since a re-numbering of the items takes place according to the space-filling curves. Additionally these routines react differently depending on the action required by the user. Since either tetrahedral, triangular, edge or nodal values have to be accessed, this is a nontrivial task. When higher order elemental unknowns are gathered, they have to be accessed from several item types, adding to the logical complexity of the gather/scatter operation.

Space-Filling Curves for Ordering Unstructured Grids

One of the unique features of `amatos` is the integrated utilization of space-filling curve (SFC) techniques for ordering and partitioning grid items. The discrete form of space-filling curves is used in computer sciences in several different situations as indicated in [429]. There are two properties of SFC that are beneficial for computations (see sect. 4.2.2):

1. mapping of d-dimensional domain to 1 dimension,
2. preservation of data locality.

In the context of `amatos`'s meshes, we construct a discrete SFC that is fine enough to have a curve-node in each element, thus inducing a consecutive numbering of elements (this is a result of the first property above). The second property guarantees connected and locally compact partitions, when the consecutive numbering is used for partitioning the computational triangulated domain.

To obtain an ordering of other mesh items like nodes or edges, we follow the ordering of elements and then collect those nodes/edges contained in the corresponding elements that have not yet been collected. This strategy results in data locality not only for elemental data but also for all other unknowns.

Interpolation of Scattered Data

For several numerical applications, interpolation of data on given grids is required. One main class of applications is the semi-Lagrangian time discretization where upstream values have to be interpolated. Since `amatos`'s grids are irregular, interpolation methods have to deal with scattered data.

There are three different types of interpolation methods available in `amatos`:

1. Bi-cubic spline interpolation based on a Hermite representation on elements;
2. radial basis function interpolation;
3. interpolation based on the finite element representation of functions.

Interpolation based on cubic splines has been described in [33, 34]. We basically estimate the gradients at element's nodes and calculate a cubic interpolating function from nodal values and nodal (estimated) derivatives. For more details on computing gradients on unstructured grids, see sect. 6.1.

The radial basis function (RBF) interpolation is described in detail in e.g. [40, 75, 217] and is covered as well in sect. 6.1.4. `amatos` uses thin plate spline RBF for interpolating scattered data. A RBF interpolation for l given function values $\rho(\mathbf{x}_i)$ at sample points $\mathbf{x}_i$, $i = 1 : l$, looks as follows

$$I_f(\mathbf{x}) = \sum_{i=1:l} [\lambda_i \cdot \phi(\|\mathbf{x}_i - \mathbf{x}\|)] + P(\mathbf{x}),$$

where $I_f(\mathbf{x}_i) = \rho(\mathbf{x}_i)$ for $i = 1 : l$ fulfills the interpolation condition, $\|\cdot\|$ is the Euclidean norm, and $P(\mathbf{x})$ is a polynomial that is required to close the arising problem of calculating coefficients λ_i from the given interpolation conditions. ϕ are the radial basis functions, in this case

$$\phi(r) = r^2 \log(r).$$

The finite element based interpolation is straight forward. Let $\mathbf{x}_i$, $i = 1 : n$ be the unknowns of the element that contains the interpolation coordinate $\mathbf{x}$. Then the interpolating function is just the FEM expansion

$$I(\mathbf{x}) = \sum_{i=1:n} \rho_i b_i(\mathbf{x}),$$

with b_i the FEM basis functions and ρ_i the corresponding FEM coefficients corresponding to nodal values $\rho(\mathbf{x}_i)$.

Note that all these interpolations are included in **amatos**'s programming interface, since interpolation needs access to individual nodes and this can be implemented much more efficiently by accessing the (object oriented) grid data directly. Moreover, the unified interface allows for rapid exchange of different orders of interpolation for testing purposes.

As an example of how to call the interpolation, we demonstrate the interpolation of upstream values in an application of atmospheric tracer transport.

Example 4.5.1 *The following code segment shows the usage of three different routines in* **amatos** *programming interface:* **grid_domaincheck**, **grid_bound-intersect**, *and* **grid_coordvalue**. *While the first two routines are used to guarantee that an upstream point in the semi-Lagrangian time stepping scheme is positioned in the domain, interpolation is invoked by* **grid_coordvalue**. *The following data structures are used:*

- **p_mesh** *is a grid handle data structure referring to the actual mesh;*
- **r_coord** *is an array with node coordinates;*
- **r_upstr** *is an array with upstream coordinates;*
- **GRID_highorder** *is pre-defined in* **amatos** *and invokes cubic spline interpolation;*
- **i_val** *is a pointer to a registered variable in* **amatos**.

We demonstrate the loop over all nodes computing the interpolated upstream value:

```
!---------- loop over nodes: find element containing upstream point
        node_loop: DO i_cnt=1, i_arlen
!---------- check if upstream value is outside of the domain
        i_out= grid_domaincheck(p_mesh(i_time), r_upstr(:,i_cnt))
!---------- take the intersection of the trajectory with the
!---------- boundary as new upstream point
        out_domain: IF(i_out /= 0) then
            r_upstr(:,i_cnt)= grid_boundintersect(p_mesh(i_time), &
                        r_coord(:,i_cnt), r_upstr(:,i_cnt), i_info=i_stat)
```

```
    no_intersect: IF(i_stat /= 0) THEN
        r_rside(i_cnt)= 0.0
        CYCLE node_loop
      END IF no_intersect
    END IF out_domain
!---------- interpolate
      r_rside(i_cnt)= grid_coordvalue(p_mesh(i_time), r_upstr(:,i_cnt), &
                      i_interpolorder=GRID_highorder, i_valpoint=i_val)
    END DO node_loop
```

Finite Element Support

amatos provides a flexible support for finite elements. The definition of element types includes the assignment of unknowns (degrees of freedom, DOFs) to grid items like elements (interior DOFs), edges (edge DOFs) and nodes. Element types are defined by a signature file.

In order to provide a flexible interface to all kinds of finite elements, amatos uses a signature data structure. The signature of a specific finite element contains the essential information of that element type:

- name, order, and total number of DOFs,
- number of DOFs on the element's vertices,
- number and position of DOFs on element's edges,
- number and position of DOFs within the element,
- coefficients of (Lagrange) basis and quadrature,
- evaluation of basis functions at quadrature points.

The signature allows for a generic programming of finite element related routines (e.g. computing of stiffness or mass matrix), independent of the basis chosen. Examples for this generic programming are given in the accompanying test driver software for amatos. With this support, it is not only possible to work with finite elements but also to apply spectral elements for high order approximations on triangles [309].

Computational Geometry Applied

Some applications related to amatos require geometric manipulations of grid items. Important classes of geometric manipulations are the point intersection of a given vector with the domain boundary, and area intersection of a given polygon with grid elements. Usually, these calculations require direct access to mesh items, so following amatos' philosophy they are included in the programming interface.

Geometric intersection of a given vector with the boundary can be computed with the routine grid_boundintersect, where a vector is given by a start and end point and the intersection point's coordinates are returned. The algorithm behind the intersection is based on the parametric representation of lines. It can be found in almost every book on computational geometry.

Algorithm 4.5.2

1. *Input: Line $\overline{AB}$ and line $\overline{XY}$;*
2. *Check whether A on same side of $\overline{XY}$ as B, if yes exit;*
3. *Check whether $\overline{AB}$ parallel to $\overline{XY}$, if yes exit;*
4. *Calculate intersection by*

$$\frac{(Y-X)\times(X-A)}{(X-Y)\times(B-A)}\cdot(B-A)+A.$$

The algorithmic realization of the area intersection can also be found in standard literature on computational geometry [9]. It follows a very simple principle, but requires both intersected polygons to be oriented and convex (convexity has to be guaranteed by the user). Given a polygon and a triangle of the triangulation. Then the algorithm runs through each edge of the triangle and cuts off those parts of the polygon that are to the right (outside) of the edge. This algorithm effectively results in calculating line intersections of the polygon lines and the triangle lines. The main ingredient is the calculation of cross products.

Algorithm 4.5.3

1. *Input: Polygon v_j $(j = 1 : n)$ and triangle $\mathbf{x}_i$ $(i = 1 : 3)$;*
2. *For each $i = 1 : 3$ DO:;*
3. *For each $j = 1 : n$ DO:*
   a) *IF v_j is right of $\overline{\mathbf{x}_i \mathbf{x}_{i+1}}$ (i.e outside of triangle) THEN*
   b) *calculate line intersection w_i of $\overline{\mathbf{x}_i \mathbf{x}_{i+1}}$ with $\overline{v_j v_{j+1}}$;*
   c) *take w_i as new node in intersection polygon.*
   d) *END IF*
4. *END DO*

Now, to intersect a given polygon with all mesh cells that have an intersection area, requires to find those cells. `amatos` uses a recursive algorithm/implementation to achieve this, accessible by the interface routine `grid_polygridintersect` (see [38] for a list of routines).

Algorithm 4.5.4

1. *Input: Polygon P and mesh M;*
2. *For each cell τ of the coarsest level in M DO:*
   a) *calculate area intersection of τ with P, $A = \tau \cap P$;*
   b) *IF $A \neq \emptyset$ AND τ has children τ_1 and τ_2 THEN:*
   c) *Set $\tau \leftarrow \tau_i$ $(i = 1 : 2)$, GO TO step 2a;*
   d) *ELSE return.*
   e) *END IF*
3. *END DO*

Structure of amatos Application

To finalize this chapter we give an abstract structure of a typical application that uses amatos as a mesh handling tool. We will not specify the individual tasks but assume that there is some method available for

- computing new values on grid data (function compute);
- computing an element based refinement criterion (function criterion);
- initialize data (function initialize);
- define domain geometry (array r_poly);
- obtain initial (coarse) grid (file triang.file);
- define finite element signature (file signature.file).

Example 4.5.5 *(Fragment of an application using amatos)*

```
!---------- initialize the fem signature
      i_typ= grid_registerfemtype(signature.file)
!---------- initialize amatos data structures
      CALL grid_initialize
!---------- register variable pointer i_var
      i_var= grid_registerfemvar(i_typ)
!---------- set coarse and fine grid level
      CALL grid_setparameter(p_grid, i_coarselevel= i1, i_finelevel= i2)
!---------- define domain and create initial triangulation
      CALL grid_definegeometry(num.vertices, r_vertexarr= r_poly)
      CALL grid_createinitial(p_grid, c_filename=triang.file)
!---------- initialize data
      CALL initialize(p_grid)
!
!---------- this is the main adaptive loop
!
      adapt_loop: while(l_adapt)
!---------- gather data from grid, use values in pointer i_var
         CALL grid_getinfo(p_grid, i_femtype=i_typ, r_dofvalues=r_vals, &
                           i_arraypoint=(/i_var/))
!---------- do computations on r_vals, r_vals is overwritten with new values
         CALL compute(r_vals)
!---------- scatter computed values back
         CALL grid_putinfo(p_grid, i_femtype=i_typ, r_dofvalues=r_vals, &
                           i_arraypoint=(/i_var/))
!---------- refinement criterion based on r_vals, returns flags
         CALL criterion(r_vals,i_flag)
!---------- apply flags to the grid
         CALL grid_putinfo(p_grid, i_elementstatus=i_flag)
!---------- adapt the grid according to flags
         CALL grid_adapt(p_grid,l_adapt)
      END DO node_loop
!---------- terminate application
      CALL grid_terminate
```

The example shows a very simple structure of an amatos-based program. After initialization that comprises registration of FEM type, variables and geometry definition, the main adaptation loop consist of only a few steps. First data is gathered from the grid by **grid_getinfo**. Then some user specified computations take place. The data has to be written back to the grid before any adaptation in order to maintain consistency of data. Flags for refinement or coarsening have to be derived from the refinement criterion and scattered into the grid data structure. A simple call to **grid_adapt** invokes refinement and coarsening and leads to a new conforming mesh.

5

Issues in Parallelization of Irregularly Structured Problems

Problems in atmospheric modeling were counted among the grand challenges a few years ago. For almost a decade there has been extensive research in order to develop modeling software that is capable of utilizing modern high performance computing architectures (see e.g. [3, 103, 118, 154, 176, 181, 186, 201, 235, 237, 302, 333, 342, 418]). High performance inevitably means parallel computing. When dealing with adaptive mesh refinement, parallelization becomes a non-trivial issue for several reasons.

- Load balancing has to be performed dynamically, during the simulation, since load imbalances can (and will) occur frequently, when local phenomena change their location.
- Adaptively refined grids are not topologically simple, therefore partitioning strategies have to be more sophisticated than for non-adaptive grids.
- Since partitioning has to take place during the simulation, it needs to be parallelizable, scalable, and fast/efficient to be useful.
- Data movement has to be minimized even with changing locations of refinement.
- Boundaries between processor's partitions have to be as short as possible to minimize inter-process communication. In other words, one is interested in maximizing the volume to surface ratio of partitions.
- With irregular (e.g. triangular) meshes it becomes difficult to maintain halo regions, i.e. overlapping redundant boundary regions for lower communication costs.

Most of the above problems can be tackled by advanced data structures as discussed in chap. 4. However, a partitioning strategy is still required. This chapter will therefore focus on partitioning methods. Load balancing for the partitions is essential. Therefore, the load balancing problem will be dealt with in more detail. We will only investigate in more detail load balancing of the computational grid. It should be noted that load balancing in adaptive methods has to consider the (hierarchical) refinement [396].

For parallel strategies to generate unstructured meshes for adaptive mesh refinement there is a rich literature. We can only mention a selection here. For some general approaches and data structures for irregular tetrahedral meshes see [173, 225, 227]. An error estimate driven domain decomposition method that uses parallel instances of serial grid generation software has been proposed by Bank and Holst [21]. Finally, it should be noted (however, not elaborated) that the efficiency of adaptive applications is severely dependent on programming models on the target computing architectures [364].

5.1 Partitioning Strategies

In order to distribute the computational load uniformly among processors, in adaptive methods it is not sufficient to just split the domain naively. Since refinement regions may occur and vanish and also may move, partitioning has to take place frequently. The partitioning strategy should lead to a distribution of mesh atoms such that the above mentioned quality criteria are met, i.e.

- equilibrate the load,
- minimize the edge cut, and
- minimize the re-partitioning cost.

Solving this problem exactly is NP-hard, therefore heuristics or approximate methods have to be employed. Additionally, the partitioning problem is in most cases an unstructured problem. In this section, we will introduce four different classes of partitioners.

- Spectral bisection partitioning;
- Kernighan-Lin partitioning;
- Multi-level methods;
- Space-filling curves.

Each such procedure is in principle capable of handling quadrilateral as well as triangular meshes. Of course there are more strategies to partitioning. The reader is referred to [62, 76, 108, 123, 124, 150, 250, 404]. An overview of parallel load balancing schemes for adaptive methods can be found in [224].

To start the discussion of (unstructured) grid partitioning, we introduce the notion of graphs. The meshes described in sect. 3 can all be represented by an undirected graph.

Definition 5.1.1 *(Undirected Graph, vertex degree)*
Let V denote the set of graph vertices, and E the set of edges. $|\cdot|$ denotes the number of set members. Then by $G(V, E)$ we denote the graph of vertices $v_i \in V$ connected by edges $e(v_i, v_j) \in E$.

Let v be a vertex of the graph. Then the degree g of the vertex is defined as $|e_n|$, where $e_n \subset E$ is the subset of E of all edges that contain v.

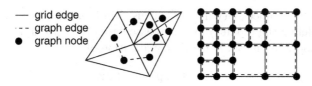

— grid edge
--- graph edge
● graph node

Fig. 5.1. Examples of graphs related to meshes. (**Left**) An element-based graph for triangular finite element meshes; (**right**) a node-based graph for quadrilateral finite difference meshes

Note that we assume implicitly that undirected graphs do not contain *self edges*, i.e. edges $e(v_i, v_i)$, and do not contain multiple edges between two vertices. We will also use the term *node* instead of vertex.

Example 5.1.2 *(Triangular 2D finite element mesh)*

For a finite element mesh it is most convenient to identify the elements with graph vertices. Then, the graph edges can be identified with the mesh edges. Even if the mesh edges (faces) separate two elements, they provide the neighborhood information required for the the graph edges to connect two elements (see fig. 5.1). The graph edges, therefore, represent the neighborhood relations in the mesh.

Example 5.1.3 *(Quadrilateral 2D finite difference mesh)*

In a finite difference mesh the graph edges correspond to the mesh edges, since the vertices correspond to the grid nodes (see fig. 5.1). The graph edges, like in example 5.1.2, represent the neighborhood relations, in other words the finite difference stencil.

Now, we are able to formulate the partitioning problem in terms of a graph. Let p be the number of processors, and let $G(V, E)$ be the graph corresponding to the given (adaptively refined) mesh, with $n = |V|$ the number of grid atoms (graph vertices) to be partitioned. Then we look for a partitioning $P = (V_1, \ldots, V_p)$ into sub-graphs $G(V_k, E_k)$ $(k = 1 : p)$ with the following properties:

1. $V_k \cap V_l = \emptyset$ for $k \neq l$, $k, l = 1 : p$.
2. $\bigcup_{k=1:p} V_k = V$.
3. $|V_k| = \frac{n}{p}$ for all $k = 1 : p$.

In most cases, we will not be able to fulfil the last property exactly, since we need an integer number of vertices in each V_k, so we might relax this to $|V_k| = \lceil \frac{n}{p} \rceil$, where $\lceil \xi \rceil = \min_{i \in \mathbb{Z}} (i \geq \xi)$ for $\xi \in \mathbb{R}$ is the *ceiling function*.

Note that the above properties imply that the partitions $G(V_k, E_k)$ are connected graphs, inducing connected sub-meshes. If this cannot be guaranteed by the partitioning method, we might relax the properties, by allowing $k = 1 : q$, with $q > p$.

In sect. 4.2 a connectivity matrix was used in illustrations. Here, we want to formally define the (sparse) matrices associated to a graph, and therefore, to the corresponding mesh. With this association of

$$\text{Mesh} \leftrightarrow \text{Graph} \leftrightarrow \text{Matrix}$$

we are able to use powerful mathematical tools for matrix analysis in order to derive appropriate partitions.

Definition 5.1.4 *(Incidence/ Laplacian/ connectivity matrix)*
    Let $G(V, E)$ be an undirected graph and let $|V| = n$, $|E| = m$. Then the incidence matrix $I \in \mathbb{N}^{n \times m}$ is defined by

$$I_{i,j} = \begin{cases} 1, & \text{if } e(v_i, v_k) \text{ is the } j - \text{th edge,} \\ -1, & \text{if } e(v_k, v_i) \text{ is the } j - \text{th edge,} \\ 0, & \text{otherwise,} \end{cases}$$

for some $1 \leq k \leq n$. The connectivity matrix $C \in \mathbb{N}^{n \times n}$ *corresponding to $G(V, E)$ is defined by*

$$C_{i,j} = \begin{cases} 1, \text{ if } i = j, \\ 1, \text{ if } e(v_i, v_j) \text{ exists,} \\ 0, \text{ otherwise.} \end{cases}$$

The Laplacian matrix $L \in \mathbb{N}^{n \times n}$ *is defined by*

$$L_{i,j} = \begin{cases} d, & \text{if } i = j, \ d \text{ degree of node } i \\ -1, & \text{if } e(v_i, v_j) \text{ exists,} \\ 0, & \text{otherwise.} \end{cases}$$

Note that the Laplacian matrix and the connectivity matrix, defined above, have the same sparsity pattern. Some authors define the diagonal of the connectivity matrix to be zero. Then one can write $L = D - C$, where D is the diagonal matrix of all vertex degrees. Note that the definition of the incidence matrix is somewhat ambiguous, since we consider undirected graphs. Therefore, each row of the incidence matrix is defined uniquely up to the sign.

    For simplicity, we will restrict ourselves in the following introduction to different partitioning methods to $p = 2$, with partitions of V into the vertex sets V_1, V_2.

5.1.1 Spectral Bisection Partitioning

Spectral bisection partitioning has been motivated and introduced by Fiedler [147, 148]. In the context of sparse matrix ordering, spectral bisection has been utilized by Pothen et al. [325]. The algorithm is based on properties that can be proved for the matrices, defined in definition 5.1.4 (for the proofs, see textbooks on algebraic graph theory, e.g. [60]).

Proposition 5.1.5 *(Properties of the matrices)*

1. $L = II^T$ *is a symmetric matrix.*
2. *From 1 follows that all eigenvalues of L are real: $\lambda_1, \ldots, \lambda_n \in \mathbb{R}$. Furthermore the eigenvectors are real and orthogonal.*
3. *If $\mathbf{e} = (1, \ldots, 1)^T$, then $L\mathbf{e} = 0$.*
4. *For $0 \neq \mathbf{b} \in \mathbb{R}^n$ eigenvector of L, i.e. $L\mathbf{b} = \lambda\mathbf{b}$, $\lambda \in \mathbb{R}$, we have*

$$\lambda = \frac{\left\| I^T \mathbf{b} \right\|_2^2}{\left\| \mathbf{b} \right\|_2^2} = \frac{\sum_{e(v_i, v_j) \in E} (b_i - b_j)^2}{\sum_i b_i^2},$$

where $\|\mathbf{b}\|_2^2 = \sum_i b_i^2$ for $\mathbf{b} \in \mathbb{R}^n$.
5. *The eigenvalues of L are non-negative, $0 \leq \lambda_1 \leq \lambda_2 \leq \cdots \leq \lambda_n$.*
6. *From 3 follows that $\lambda_1 = 0$. The multiplicity of the trivial eigenvalue is equal to the number of connected components of the graph.*
7. *If the graph is connected, then $\lambda_2 > 0$.*

Note that the second smallest eigenvalue λ_2 is called *algebraic connectivity* of the graph $G(V, E)$.

Using these properties the algorithm of spectral bisection is simply

Algorithm 5.1.6 *(Spectral bisection partitioning)*

1. *Compute the eigenvector $\mathbf{b}^{(2)} = (b_1^{(2)}, \ldots, b_n^{(2)})$ corresponding to the algebraic connectivity λ_2.*
2. **FOR** *each vertex $v_i \in V$* **DO:**
   a) **IF** $b_i^{(2)} < 0$ *put vertex v_i in partition V_1.*
   b) **ELSE** *put vertex v_i in partition V_2.*
   c) **END IF**
3. **END FOR**

Fiedler proves that if $G(V, E)$ is connected and V_1 and V_2 are partitions obtained by algorithm 5.1.6 then V_1 is connected. Furthermore, if $b_i^{(2)} \neq 0$ for all $i = 1 : n$, then also V_2 is connected.

Since we have to compute a specific eigenvalue-eigenvector pair, the spectral bisection algorithm is hardly ever used for large meshes. In most cases it is combined with a multi-level approach as described in subsect. 5.1.3.

5.1.2 Kernighan-Lin Partitioning

The Kernighan-Lin heuristic minimizes the edge cut. It was motivated by electronic circuit design problems [238]. We have to introduce the *edge weight* first.

Definition 5.1.7 *(Edge weight)*

Let $e(v_i, v_j) \in E$ be an edge in the graph $G(V, E)$. Then we associate an edge weight $\omega(v_i, v_j) \in \mathbb{R}$ to the edge.

Note that with this definition, we can reformulate the edge cut (see appendix B) more formally:

Definition 5.1.8 *(Edge cut revisited)*
    Let $G(V, E)$ be a graph with a vertex set $V = V_1 \cup V_2$ partitioned into two subsets. Then the cut weight $W_{V_1 \cap V_2}$ *can be defined by*

$$W_{V_1 \cap V_2} = \sum_{v_i \in V_1, v_j \in V_2} \omega(v_i, v_j).$$

With this and the total weight $W = \sum_{e(v_i, v_j) \in E} \omega(v_i, v_j)$, the (weighted) edge cut *can be defined by*

$$\mathcal{E}_\omega = \frac{W_{V_1 \cap V_2}}{W}.$$

The Kernighan-Lin heuristic tries to minimize $\mathcal{E}_\omega$, resp. $W_{V_1 \cap V_2}$ by means of a greedy algorithm.

Algorithm 5.1.9 *(Kernighan-Lin algorithm)*

1. *Compute a random partition $V_1 \cup V_2 = V$.*
2. FOR *all edges $e(v_i, v_j)$ with $v_i \in V_1$ and $v_j \in V_2$* DO:
    a) *Exchange vertices v_i and v_j virtually (i.e. put vertex v_i into V_2 and v_j into V_1).*
    b) *Compute new $W_{V_1 \cap V_2}$.*
3. END DO
4. *Select the pair (v_i, v_j) with the lowest value for $W_{V_1 \cap V_2}$ and obtain new partition $V_1 \cup V_2 = V$.*
5. GOTO *step 2 and repeat the procedure until no further gain is obtained.*

Note that the given algorithm is equally valid for graphs with unit weights, i.e. un-weighted graphs. The Kernighan-Lin algorithm has computational complexity of order $\mathcal{O}(n^3)$, where $n = |V|$. Therefore, it is hardly used for large meshes/graphs, but most commonly combined with multi-level methods.

5.1.3 Multi-Level Methods

Multi-level methods follow a simple and widely used principle: Do the bulk work on a coarse sub-set of the global problem, that represents the main features, and then recursively refine the result on larger and larger sub-sets, computing only corrections to the coarse solution. The gain in efficiency results from the fact that corrections to a given (even poor) solution can often be computed in much less time than the solution itself.

We follow the original description of a multi-level graph partitioning algorithm by Hendrickson and Leland [200], also used e.g. in [233] and [352]. We first sketch the basic algorithm and then describe the single steps.

Algorithm 5.1.10 *(Multi-level graph partitioning)*

1. *Start with $k = 0$, $G(V^{(k)}, E^{(k)}) = G(V, K)$.*
2. **WHILE** *graph is too large,* **DO**
    a) *coarsen graph:* $k := k+1$, $G(V^{(k)}, E^{(k)}) \leftarrow \text{coarsen}[G(V^{(k-1)}, E^{(k-1)})]$
3. **END WHILE**
4. *On coarsest level ($k = k^\star$), partition graph into two partitions $V^{(k^\star)} = V_1^{(k^\star)} \cup V_2^{(k^\star)}$.*
5. **WHILE** *refined graph is not original graph, i.e. $V^{(k)} = V_1^{(k)} \cup V_2^{(k)} \neq V$,* **DO:**
    a) *refine the graph, i.e. ($k := k - 1$, $l = 1, 2$) $\hat{V}_l^{(k-1)} \leftarrow \text{uncoarsen}[V_l^{(k)}]$.*
    b) *optimize partition boundaries, $V_l^{(k-1)} \leftarrow \text{optimize}[\hat{V}_l^{(k-1)}]$.*
6. **END WHILE**

The coarsening of a graph can be achieved by first finding a *maximal matching*, that is a maximal set of edges, such that no two of them are incident on the same node. Then the nodes, corresponding to the edges in the maximal matching are collapsed. The weights of collapsed nodes are summed, while edges retain their weights, unless they are adjacent, i.e. there are two edges that connect each of the two collapsed nodes to the same third node. Then the edge weights are summed as well. In the coarsening step, each node in the fine graph is uniquely mapped to a node in the coarse graph. Therefore, a partition of the coarse graph induces a unique partition of the fine graph. Furthermore, a coarse graph has the same local properties regarding edge cut/weight and node weight as the fine (original) graph.

Since the refined partitions have more degrees of freedom (i.e. more vertices), the coarse partition might not be the optimum partition for this level of refinement. Therefore, in the refinement step, a further (local) optimization of partition boundaries improves the partition quality.

Note that this algorithm can be implemented with $\mathcal{O}(n)$ complexity. Furthermore, this algorithm can be parallelized as shown by Karypis and Kumar [232, 233]. So most of the current state of the art mesh partitioners use a multi-level approach [199, 234, 312, 326, 403].

5.1.4 Space-Filling Curves

A special space-filling curve (SFC) for triangular and tetrahedral meshes induced by bisection refinement, has been introduced by algorithm 4.2.5 in sect. 4.2. There is a large variety of space-filling curves in use. An overview of SFC is given in [411, 429]. Other applications of SFC in parallelizing adaptively refined meshes are given in [51, 116, 179, 322].

The basic algorithm consists of only 3 steps:

Algorithm 5.1.11 *(Space-filling curve partitioning)*

1. *Construct a SFC such that each graph vertex contains at least one point of the (discrete) SFC.*

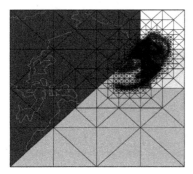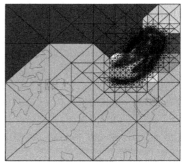

Fig. 5.2. Examples of partitioned adaptively refined meshes: Partitioning for eight processors, each color represents a processor. (**Left**) Partitioning with SFC; (**right**) with multi-level method

2. *Renumber the graph vertices along the SFC consecutively.*
3. *Partition the thus created index set into p equal sized sets.*

We are not going to introduce the diverse algorithms for creating a space-filling curve for a given graph. Sierpinsky's curves have proven to be adequate for triangular meshes (resp. the corresponding graphs), while Hilbert curves or Morton curves proved to be advantageous for quadrilateral elements. Zimmermann shows that Hilbert's curve can be easily extended to 3D [203, 429]. Most modern algorithms for creating space-filling curves work with recursively altering bit patterns. The creation of SFC in most cases can be accomplished very efficiently with parallelizable codes.

Note that the load balance for SFC induced partitions is almost optimal by construction, since the partition is constructed by cutting the 1D consecutive index set into equal sized sub-sets. Since SFC have a neighborhood preserving property, the partitions are connected for most types of SFC. Finally the edge cut is not optimal, however, it can be shown that the edge cut remains within a limited bounded neighborhood of the optimal edge cut [213, 430].

In fig. 5.2 we show the result of a multi-level partitioning and a SFC partitioning of a mesh, obtained from an application of tracer transport in the arctic stratosphere [39]. The SFC partitioning has been obtained by the built in SFC ordering in `amatos` (see sect. 4.5). The multi-level partitioning has been obtained by applying Metis, a state of the art graph partitioner [231]. A more thorough comparison of the results can be found in [45].

6

Numerical Treatment of Differential Operators on Adaptive Grids

This chapter is concerned with the realization of differential operators, mainly on unstructured and nonuniform grids. While a finite difference approximation to a differential operator can be easily derived for orthogonal and quadrilateral grids, it is not straight forward to do the same for unstructured and non-orthogonal grids. We will focus the presentation on the gradient operator

$$\nabla \rho = \left(\frac{\partial \rho}{\partial x_1}, \ldots, \frac{\partial \rho}{\partial x_d} \right), \tag{6.1}$$

where $\mathbf{x} = (x_1, \ldots, x_d) \in \mathbb{R}^d$ and $\rho : \mathbb{R}^d \to \mathbb{R}$ is a scalar function. We restrict ourselves to the gradient operator, since all of the other operators used in this book can be derived from the gradient:

$$\Delta \rho = \nabla \cdot (\nabla \rho) \quad \text{Laplace's operator,}$$
$$\text{rot}(\mathbf{u}) = \nabla \times \mathbf{v} \quad \text{curl operator,}$$
$$\text{div}(\mathbf{u}) = \nabla \cdot \mathbf{v} \quad \text{divergence operator,}$$

where ρ is a scalar and $\mathbf{u}$ a vector valued function, which is sufficiently regular to be differentiated in the above given way.

6.1 Approximating the Gradient

There are different methods to approximate a gradient operator. In this section we will introduce four different approaches to compute a discrete gradient. Note that these are all numerical gradient approximations.

- Finite difference approach;
- Least squares approximation;
- Galerkin approach;
- Radial basis function interpolatory approach.

We do not consider methods of automatic or algorithmic differentiation. For an introduction and description of current technological knowledge of automatic differentiation techniques the reader is referred to [180].

6.1.1 Finite Difference Approach

On quadrilateral and orthogonal grids, it is easy to discretize the gradient component-wise by a finite difference formula:

$$\partial_{x_i}\rho = \frac{\partial \rho}{\partial x_i} \approx \frac{\rho(x_i) - \rho(x_i - \Delta x_i)}{\Delta x_i}, \qquad (6.2)$$

where Δx_i is a step size in the x_i-coordinate direction. It is well known that this (backward) finite difference approximation is first order accurate and there are many higher order finite difference schemes available (see e.g. [288, 370]).

For unstructured grids, this approach has to be modified. There are two already quite old methods, that will be reviewed here, one proposed by Klucewicz [244] and implemented by Renka [337], the other one proposed by Akima [5]. All the mentioned references are originally concerned with interpolation of irregularly spaced data. However, the interpolation procedures rely on gradient data, thus the papers contain descriptions of gradient estimation. These methods have been described and tested for unstructured triangular grids in [33], so we only briefly introduce them here. Note that a more general approach to representing differential operators on unstructured meshes can be found in [356].

Klucewicz's method [244] is based on a mean of all gradients in grid cells surrounding the corresponding node. So for locally adapted meshes the algorithm in two dimensions is given by

Algorithm 6.1.1 *(Gradient estimation according to Klucewicz)*
    Let $\mathbf{x}_i = (x_i, y_i) \in \mathbb{R}^2$ *be the node at which a gradient is to be calculated.*
Let

$$I_i = \{(j,k) : j \neq k \neq i, \text{ such that } (\mathbf{x}_i, \mathbf{x}_j, \mathbf{x}_k) \text{ form a triangle}\}$$

be the index set of node indices in the patch of node $\mathbf{x}_i$.

1. *Calculate the plane that represents the slope in each triangle defined by* I_i.
   The plane is given by the equation

$$p_{ijk}(x, y) = \rho(\mathbf{x}_i) + c_{ijk}^{(x)}(x - x_i) + c_{ijk}^{(y)}(y - y_i), \qquad (6.3)$$

   where

$$c_{ijk}^{(x)} = \frac{\partial p_{ijk}}{\partial x}, \quad c_{ijk}^{(y)} = \frac{\partial p_{ijk}}{\partial y}$$

   have to be chosen such that

$$p_{ijk}(x_j, y_j) = \rho(\mathbf{x}_j), \quad p_{ijk}(x_k, y_k) = \rho(\mathbf{x}_k).$$

2. *Form the convex combination*

$$\overline{p_i(x,y)} = \frac{1}{w_i} \sum_{j,k \in I_i} w_{ijk} p_{ijk}(x,y)$$

with $w_i = \sum_{j,k \in I_i} w_{ijk}$, w_{ijk} *weights.*
3. *The gradient is then given by*

$$\nabla \rho|_x(\mathbf{x}_i) = \frac{1}{w_i} \sum_{j,k \in I_i} w_{ijk} c_{ijk}^{(x)}, \quad \nabla \rho|_y(\mathbf{x}_i) = \frac{1}{w_i} \sum_{j,k \in I_i} w_{ijk} c_{ijk}^{(y)}.$$

Note that algorithm 6.1.1 is based on the finite difference approximation of the gradient components in (6.3). Since in 3D a similar representation of a hyperplane can be given, an extension to three dimensional settings is straight forward.

The basic idea behind the second method in this section is similar to the above one. It also computes means of surrounding slopes. However, the implementation differs, since it is based on the unit normal vectors to the planes representing such slopes. It turns out that this method is slightly advantageous compared to Klucewicz's method, so it is selected for the comparison of different methods in sect. 6.2, denoted by FDM.

Algorithm 6.1.2 *(Gradient estimation according to Akima)*
Let $\mathbf{x}_i$ and I_i be defined as in algorithm 6.1.1.

1. *For each element in the patch of node* $\mathbf{x}_i$*, defined by* I_i*, compute the vectors*

$$\xi_j = (\mathbf{x}_j - \mathbf{x}_i), \text{ and } \xi_k = (\mathbf{x}_k - \mathbf{x}_i).$$

Furthermore, compute the vector product $\xi_{j,k} = \xi_j \times \xi_k$*.*
2. *Compute the mean* χ *of the* $\xi_{j,k}$ *over all* N *patch elements*

$$\chi = \frac{1}{N} \sum_{j,k \in I_i} \xi_{j,k}.$$

3. *Normalize* χ *such that its third component is non-negative.* $\chi = (\chi_1, \chi_2, \chi_3)$
is perpendicular to the plane defined by the slope of ρ *in node* i*, given by the plane equation*

$$\widehat{p_i(x,y)} = \frac{\chi_1}{\chi_3} x - \frac{\chi_2}{\chi_3} y - C,$$

where C *is a constant chosen suitably. Thus, the gradient is given by*

$$\nabla \rho|_x(\mathbf{x}_i) = -\frac{\chi_1}{\chi_3}, \quad \nabla \rho|_y(\mathbf{x}_i) = -\frac{\chi_2}{\chi_3}.$$

6.1.2 Least Squares Approximation

In order to estimate the gradient at any position in the grid, a least squares approach seems natural. Let us consider a two-dimensional problem with a scalar function $\rho : \mathbb{R}^2 \to \mathbb{R}$ for which the gradient $\nabla \rho$ is to be estimated. Let $\{(x_i, y_i), \rho_i\} \in \mathbb{R}^2 \times \mathbb{R}$ be coordinate-value pairs representing the (discrete) function ρ. Then the aim is to find a plane (representing the gradient), given by the linear parametric representation

$$p(x, y) = ax + by + c,$$

with (a, b, c) a parameter triplet. With the least squares approach, p can be found by requiring it to be the best fit with respect to the l^2-norm:

$$\|p - \rho\|_2 = \min!$$

This can be translated into an over-determined system of linear equations $Ax = b$ with

$$A = \begin{bmatrix} x_1 & y_1 & 1 \\ x_2 & y_2 & 1 \\ & \vdots & \\ x_m & y_m & 1 \end{bmatrix}, \quad x = \begin{bmatrix} a \\ b \\ c \end{bmatrix}, \quad b = \begin{bmatrix} \rho_1 \\ \rho_2 \\ \vdots \\ \rho_m \end{bmatrix}. \tag{6.4}$$

Note that we use the Vandermonde matrix A, which for larger number of parameters becomes very ill conditioned. Solving this system of equations by a QR-algorithm, yields the three parameters $[a, b, c]^T$. Now, observing the definition of p, we see that we obtain an approximation to the gradient by

$$\nabla \rho \approx \nabla_h \rho := \nabla p = (a, b).$$

A much more elaborate approach for upwind finite difference type methods for fluid flow problems on arbitrary meshes can be found in [105].

6.1.3 Galerkin Approach

A method for solving partial differential equations that was historically developed on unstructured meshes is the finite element method. We will briefly introduce the basic principle here and derive from that principle a method to compute gradients as well.

We start with the formulation of a model problem, namely Poisson's equation on a bounded domain $\mathcal{G} \subset \mathbb{R}^d$ with boundary $\partial \mathcal{G}$:

$$-\Delta \rho = f, \text{ in } \mathcal{G}, \quad \rho = g, \text{ on} \partial \mathcal{G}. \tag{6.5}$$

For simplicity we will assume for the function $g : \partial \mathcal{G} \to \mathbb{R}$ that $g \equiv 0$, meaning that we assume homogeneous Dirichlet boundary conditions. $\rho : \overline{\mathcal{G}} \to \mathbb{R}$ is the unknown function and $f : \mathcal{G} \to \mathbb{R}$ a given right hand side. We assume suitable regularity for the functions involved without specifying the details here.

The variational formulation of (6.5) is given by

$$\text{Find } \rho \in \mathcal{V} \text{ such that}$$
$$a(\rho, b) = f(b), \text{ for all } b \in \mathcal{V}, \tag{6.6}$$

with a a bilinear form and f a linear form defined below, $b \in \mathcal{V}$ a test function or basis function and $\mathcal{V} = H_0^1(\mathcal{G})$ a suitably chosen function space, in this case the Sobolev space of compactly supported integrable functions.

$$a(\rho, b) = \int_{\mathcal{G}} \nabla\rho \, \nabla b \, dx,$$

$$f(b) = \int_{\mathcal{G}} fb \, dx.$$

The derivation of the variational form, Sobolev spaces, and existence and convergence theorems can be found in common finite element books (see e.g. [2, 71]).

The *Galerkin approach* to solving (6.6) introduces suitable basis functions $b_i \in \mathcal{V}_h$ such that ρ can be represented by the expansion

$$\rho(\mathbf{x}) = \sum_{i=1:N} \rho_i b_i(\mathbf{x}), \tag{6.7}$$

where $\rho_i \in \mathbb{R}$ are unique coefficients. In fact, $span\{b_1, \ldots, b_N\} = \mathcal{V}_h$, which means that the set $\{b_1, \ldots, b_N\}$ forms an N-dimensional basis of $\mathcal{V}_h$. Furthermore we denote by $\mathcal{V}_h$ the corresponding finite dimensional approximation to $\mathcal{V}$. Since $\mathcal{V}_h \subset \mathcal{V}$ we can now formulate the discrete form of the variational equation (6.6):

$$\text{Find coefficients } (\rho_1, \ldots, \rho_N) \in \mathbb{R}^N \text{ such that}$$
$$\sum_i a(\rho_i b_i, b_j) = f(b_j), \text{ for } i, j = 1 : N. \tag{6.8}$$

a and f are taken as defined above in the non-discrete formulation. Note that since the b_i form a basis of $\mathcal{V}_h$, we only have to test (6.8) over a finite number of basis functions. In other words, we have formulated a system of equations

$$A_h \rho_h = f_h,$$

with

$$A_h = \begin{bmatrix} a(b_1, b_1) & \cdots & a(b_1, b_N) \\ \vdots & & \vdots \\ a(b_N, b_1) & \cdots & a(b_N b_N) \end{bmatrix}, \quad \rho_h = \begin{bmatrix} \rho_1 \\ \vdots \\ \rho_N \end{bmatrix}, \quad f_h = \begin{bmatrix} f(b_1) \\ \vdots \\ f(b_N) \end{bmatrix}.$$

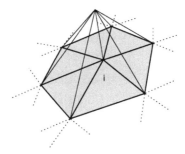

Fig. 6.1. Piecewise linear basis function b_i

The *finite element method* (FEM) constructs basis functions b_i such that A becomes a sparse matrix, by defining b_i only on a local (and small) support. This is done by introducing a triangulation of the domain $\mathcal{G}$, to obtain a polygonal approximation $\mathcal{G}_h$. Then, basis functions are defined piecewise on each element. Conforming FEM require continuity from element to element, while non-conforming FEM allow for discontinuities along cell edges. The simplest conforming finite element that can be used for solving Poisson's equation and which is continuously differentiable, consists of piecewise linear basis functions. Let $\mathbf{x}_i$, $i = 1 : N$ be nodes of a triangulation. Then we define

$$b_i(\mathbf{x}_j) = \delta_{ij},$$

where δ_{ij} is the discrete Dirac delta function, and linear in between the nodes. A sketch of this hat function is depicted in fig. 6.1.

Depending on the problem, the system of equations may be nonlinear. In the case of the above formulated model problem (Poisson's equation (6.5)) the system is linear.

To summarize this very brief outline of the finite element method, the following steps are required to realize a discretization of an equation of the form

$$\mathcal{D}(\rho) = r,$$

where $\mathcal{D}$ denotes a differential operator, ρ is the unknown function and r a known right hand side.

1. Triangulate the domain $\mathcal{G}$ with a suitable triangulation to obtain $\mathcal{G}_h$.
2. Choose appropriate function spaces $\mathcal{V}$ and $\mathcal{V}_h$ to represent the differential operator together with boundary and initial conditions in continuous and discrete space.
3. Define basis functions $b_i \in \mathcal{V}_h$ such that the discretization matrix A_h becomes sparse. In fact there are more requirements on the basis functions, like preservation of characteristic properties of the equation, that might influence the choice of the basis.
4. Set up the system of equations, defined by the bilinear form a and the linear form f and solve it.

5. Reconstruct the function ρ by the Galerkin expansion (6.7).

Once the solution vector of coefficients $(\rho_1, \ldots, \rho_N)^T$ has been computed, the function ρ can be reconstructed by means of the expansion (6.7). If we look for the gradient $\nabla\rho$ of ρ, we can use the expansion again. Apply the ∇-operator on both sides of (6.7) to get

$$\nabla\rho(x) = \sum_{i=1:N} \rho_i \nabla b_i. \tag{6.9}$$

When computing $\nabla\rho$ at nodes, based on piecewise functions, we face a problem. For example, the piecewise linear basis functions defined above and depicted in fig. 6.1 are not differentiable at cell interfaces and especially not at nodes. To solve this problem, one could define higher order basis functions with higher regularity conditions along cell interfaces. Another possibility is to compute a weighted sum or mean of local gradients:

$$\nabla\rho(\mathbf{x}_i) = \frac{1}{n} \sum_{\tau_j \in I_i} |\tau_j| \cdot \nabla\rho|_{\tau_j}(\mathbf{x}_i),$$

where $n = \sum_{\tau_j \in I_i} |\tau_j|$ is the area of the patch corresponding to node $\mathbf{x}_i$, defined by I_i (see sect. 6.1.1), τ_j are the cells, and $|\cdot|$ is the area of a cell.

6.1.4 Radial Basis Function Approach

Interpolatory methods for deriving the gradient operator on unstructured grids work according to the following simple idea:

1. Given a set of nodes $\{\mathbf{x}_1, \ldots, \mathbf{x}_l\} \subset \mathbb{R}^d$ and a set of function values corresponding to the nodes, $\{\rho_1, \ldots, \rho_l\}$, find a polynomial interpolation function $p : \mathbb{R}^d \to \mathbb{R}$ such that

$$p(\mathbf{x}_i) = \rho(\mathbf{x}_i), \quad \text{for all } i = 1 : l.$$

2. Since p is a polynomial, it is easy to derive the analytical partial derivative ∇p.
3. Then set
$$\nabla_h\rho := \nabla p.$$

This principle works with all kinds of polynomials p, but an especially interesting class of polynomials for locally unstructured grids and therefore irregularly distributed nodes $\mathbf{x}_i$ are the radial basis functions. This approach dates back to Hardy [189] and Micchelli [286]. The interpolating function p is formed by one-dimensional basis functions that are radially symmetric. Thus, an ansatz

$$p(\mathbf{x}) = \sum_{i=1:l} \lambda_i \varphi(\|\mathbf{x} - \mathbf{x}_i\|) \tag{6.10}$$

can be chosen, where λ_i are coefficients to be determined from the interpolation problem and $\varphi : \mathbb{R} \to \mathbb{R}$ is a sufficiently smooth basis function. Together with the interpolation problem

$$p(\mathbf{x}_i) = \rho_i, \quad for\ all\ i = 1 : l$$

We have to solve a $l \times l$ linear system of equations $A\lambda = \rho$. Denoting $\varphi_{ij} = \varphi(\|\mathbf{x}_i - \mathbf{x}_j\|)$ we have

$$A = \begin{bmatrix} \varphi_{11} & \cdots & \varphi_{1l} \\ \vdots & & \vdots \\ \varphi_{l1} & \cdots & \varphi_{ll} \end{bmatrix}, \quad \lambda = \begin{bmatrix} \lambda_1 \\ \vdots \\ \lambda_l \end{bmatrix}, \quad \rho = \begin{bmatrix} \rho_1 \\ \vdots \\ \rho_l \end{bmatrix}. \tag{6.11}$$

Depending on the particular basis function, sometimes we have to add a polynomial part to the ansatz (6.10), in order to be able to invert A in (6.11). This means, we define p by

$$p(\mathbf{x}) = \sum_{i=1:l} \lambda_i \varphi(\|\mathbf{x} - \mathbf{x}_i\|) + q(\mathbf{x}).$$

To be solvable, we restrict the coefficients to fulfill

$$\sum_{i=1:q} \lambda_i p(\mathbf{x}_i) = 0.$$

From the interpolation problem and the above restriction it follows, that we now solve a blocked system $B\Lambda = \Xi$, where

$$B = \begin{bmatrix} A & P \\ P^T & 0 \end{bmatrix}, \quad \Lambda = \begin{bmatrix} \lambda \\ \mu \end{bmatrix}, \quad \Xi = \begin{bmatrix} \rho \\ 0 \end{bmatrix}, \tag{6.12}$$

where μ is the vector of coefficients or the polynomial q,

$$q(\mathbf{x}) = \sum_{i=1:r} \mu_i \mathbf{x}^{i-1},$$

and P is formed by evaluations of $q(\mathbf{x}_i)$. Table 6.1 lists some common radial basis functions together with the order r of polynomial q. The algorithm for computing the gradient operator at any given point is finally given by

Algorithm 6.1.3 *(Radial basis function interpolatory gradient calculation)*
   Let $\mathbf{x}$ *be a given coordinate,* $\{\mathbf{x}_1, \ldots, \mathbf{x}_l\}$ *be the set of nearest mesh points with corresponding values* $\rho_i,\ i = 1 : l$.

1. *Evaluate* $\varphi(\|\mathbf{x} - \mathbf{x}_i\|)$ *for* $i = 1 : l$ *in order to derive the coefficients of* A, *and evaluate* $q(\mathbf{x}_i)$ *for setting up* P, *if necessary.*
2. *Solve the linear system of equations given by either (6.11) or (6.12).*

Table 6.1. Common radial basis functions with their corresponding polynomial expansion order

name	formula	order of q
thin plate splines	$\varphi(r) = r^2 \log r$	$r = 2$
multi-quadrics	$\varphi(r) = \sqrt{(r^2 + 1)}$	$r = 1$
Gaussians	$\varphi(r) = \exp(-r^2)$	$r = 0$
inverse multi-quadrics	$\varphi(r) = (r^2 + 1)^{-1/2}$	$r = 0$

3. *Evaluate* $\nabla p(\mathbf{x})$, *where*

$$\nabla p = \sum \varphi' + \nabla q.$$

Note that we have to solve a (small) system of equations for each set of neighboring grid points $\{\mathbf{x}_1, \ldots, \mathbf{x}_l\}$. Since the radial basis functions are only dependent on the distance of mesh point $\mathbf{x}_i$ to the given coordinate $\mathbf{x}$, there is no mesh required in order for the calculation to be well defined. For more details on radial basis functions in general, the reader is referred to [75]. Fornberg and Flyer study the accuracy of radial basis functions for derivative estimation [153]. For radial basis functions in advection problems, see [40, 41]. Hubbert has studied interpolation problems by radial basis function approaches on the sphere [208, 209]. Another study of scattered data interpolation on spheres can be found in [299]. A compactly supported RBF approach is taken to solve the shallow water equations in [417].

6.2 Evaluating the Gradient Approximation

In order to evaluate the approximations derived in the previous section, we utilize a two-dimensional test case. Let

$$\rho(x, y) = \cos\left(2\pi(x - \frac{1}{4})\right) \cos\left(2\pi(y - \frac{1}{4})\right)$$

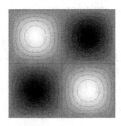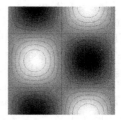

Fig. 6.2. The basic test problem for gradient computation: (**left**) the scalar function ρ, (**center**) $\nabla\rho|_x$, and (**right**) $\nabla\rho|_y$

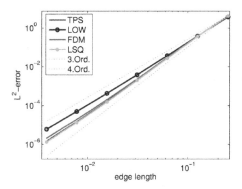

Fig. 6.3. Convergence of the gradient approximation methods. The dotted lines represent third and fourth order ideal convergence rates

be given on the two-dimensional unit square $[0,1]^2 \subset \mathbb{R}^2$. A plot of this function together with the two gradient components is given in fig. 6.2. The gradient of this function can be given analytically by

$$\nabla\rho|_x = -2\pi \sin\left(2\pi(x - \frac{1}{4})\right)\cos\left(2\pi(y - \frac{1}{4})\right),$$

$$\nabla\rho|_y = -2\pi \cos\left(2\pi(x - \frac{1}{4})\right)\sin\left(2\pi(y - \frac{1}{4})\right).$$

In order to evaluate some of the methods, described in the previous sections, we compare the numerically computed gradient $\nabla_h\rho$ with the exact one, $\nabla\rho$ in the l^2-norm (N the number of grid points):

$$\varepsilon_{l^2} = \|\nabla_h\rho - \nabla\rho\|_2 = \frac{1}{N}\left(\sum_{i=1:N}(\nabla_h\rho(x_i,y_i) - \nabla\rho(x_i,y_i))^2\right)^{\frac{1}{2}}.$$

Figure 6.3 shows the convergence behavior of the different methods. We denote the thin plate spline interpolatory gradient approximation by *TPS*, the finite difference based method by *FDM*, a linear finite element method based Galerkin approach by *LOW*, and the least squares based method by *LSQ*. Figure 6.3 shows that the LSQ method on a uniform grid performs best, together with the TPS method.

When looking at the difference plots (fig. 6.4) we see that the largest (and only visible) error occurs at the boundary of the domain. This is not surprising, since all methods rely on collecting surrounding node/element values in order to approximate the gradient operator. Since at the boundary, only node/element values on one side of the corresponding position exist, accuracy drops drastically.

Next, we investigate the dependency of the corresponding method to local irregularities in the mesh. It is of great importance to keep these effects limited,

Fig. 6.4. Error plots for four different gradient approximation methods on a uniform mesh (CNT[15], corresponding to $h \approx 5.5 \cdot 10^{-3}$. From left to right: TPS, FDM, LOW, LSQ. Differences are hardly visible and are limited to the boundary

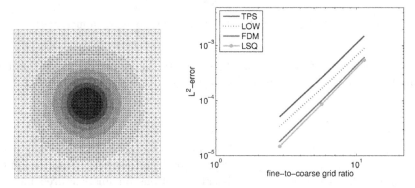

Fig. 6.5. Dependence on non-uniformity of the mesh: (**left**) a locally refined mesh (the CNT[11 : 17] case); (**right**) l^2-norm of the error for cases CNT[11 : 17], CNT[13 : 17], CNT[15 : 17], resp. (mesh ratio is given)

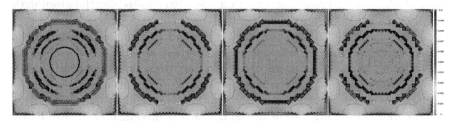

Fig. 6.6. Error plots for four different gradient approximation methods(like fig. 6.4) on a adaptively refined mesh (see fig. 6.5 for the mesh)

as shown e.g. in [131, 184]. In fact, all methods show dependence on the mesh geometry. Surprisingly the TPS method is affected most. Looking at the difference plots (fig. 6.6), we see that the largest error occurs at grid interfaces.

Finally, the computational effort has to be considered. The TPS method, although not numerically exceptionally advantageous, is by a factor of 4 to 5 more costly than the LOW method, while the LSQ method is a factor of 1.5 to 2 more expensive than the LOW method. However, all three methods show

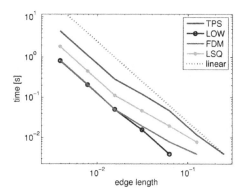

Fig. 6.7. Timing results for the gradient approximation methods

perfect linear order of computational complexity, i.e. for increasing the number of mesh points, the computational cost grows linearly with this number.

6.2.1 Further References

For a better representation of gradients on unstructured meshes, it might be helpful to look at different approaches. Abgrall and Harten introduce multi-level approaches to represent multi-scale functions on unstructured meshes [1]. An old and successful method is the combination of finite element theory with least squares approximation [149]. Since the thin plate spline radial basis function approach in our examples did not perform too satisfactorily, one could think of using other basis functions that are better suited for the given data [75, 217]. Radial basis functions have been used successfully to recover vector valued functions [29]. A rapid evaluation method that further minimizes the effort of solving a dense system of equations for each set of interpolation points can be found in [348]. A stencil based approach that considers data locality and is suitable for parallelization is proposed in [59] and in a different way in [133]. Interpolatory methods might profit from insight in the spatial error of the underlying interpolation method, which is investigated in [79]. Kirby and Karniadakis introduce de-aliasing techniques in order to improve results of continuous and discontinuous Galerkin methods on non-uniform grids [241]. A stable variational approach for numerical differentiation is presented in [245].

7
Discretization of Conservation Laws

Many of the basic equations in atmospheric modeling are based on conservation laws. Conservation of mass constitutes the continuity equation, and conservation of momentum establishes the momentum equations. When conservation properties are present in the continuous equations, the numerical (discrete) counterparts should also have conservative properties. Examples for numerical conservation of vorticity or other state variables can be found in [46, 350]. More generally we want a numerical method to adhere to a *structure preservation property*. To achieve conservation is paramount for all kinds of numerical methods that try to discretize conservation laws. However, for adaptive methods this often poses an additional challenge.

In this chapter we first introduce some of the basic mathematical tools like the *divergence theorem* and others. We will then deal with deriving the basic systems of equations used in atmospheric modeling from conservation principles. Once these equations have been established, we present several methods to solve these. Our main emphasis will be on semi-Lagrangian methods in sect. 7.4. Two other sections are devoted to the finite volume method (sect. 7.2) and the discontinuous Galerkin method (sect. 7.3).

There are of course many more approaches to achieve conservation in adaptive flow problems. A survey of (mostly geosciences related) advection methods can be found in [141]. Mixed-hybrid finite element methods achieve conservation for fluid flow. An overview can be found in [145]. For a projection method in the context of adaptive mesh refinement the reader is referred to [6]. A modified method of characteristics Galerkin scheme with local conservation properties has been proposed in [10]. Spectral methods for hyperbolic conservation laws are reviewed in [175]. A comparison of different finite element methods for solving the shallow water equations is given in [188]. High order finite differences are employed in [420], while a positive advection method for tracer transport on unstructured grids is introduced in [413]. An investigation of conservation of smoothed particle hydrodynamics for the shallow water equations can be found in [155].

In this work we will not go into the very details of geophysical fluid flow. However, most geophysical flow problems comprise multiple time scales. In many cases, fast modes are not considered or filtered in order to maintain numerical stability. A comparison of several such methods, is given in [27]. More generally an overview of unstructured grid techniques in CFD can be found in [278]. In sect. 7.4, we consider semi-implicit methods. For further evaluation of implicit schemes see [48, 49, 50, 261, 293, 335]. In order to integrate (semi-) implicit schemes, large systems of equations have to be solved. We will not consider to present solution techniques here, since this would go beyond the scope of this book. However, we refer to [65, 66, 68, 82, 127, 144, 156, 202, 239, 422] among others.

7.1 Conservation Laws of Interest

In this section we derive some of the common equations of atmospheric modeling from basic conservation laws. Especially in climate modeling, it is essential to preserve these conservation properties in the discretization. Our presentation will be restricted to conservation of mass, momentum, and energy and we will restrict ourselves to the very basic principles.

It should be noted that even if we pretend momentum to be a conservation quantity, the equation derived in subsect. 7.1.2 strictly speaking is not a conservation principle. Additionally, in meteorological applications the conservation of *angular momentum* is often much more important and can be stated as a strict conservation principle. We refer the reader to the dynamical meteorology literature for a more comprehensive presentation of the derivation of the main equations [134, 165, 311].

7.1.1 Conservation of Mass

The most basic conservation principle is *conservation of mass*. We are going to derive the continuity equation from this principle in a one-dimensional setting. This example corresponds to fluid flow through a tube, where we assume the fluid to be constant across each cross section of the tube. Let x represent the space coordinate, and t the time. $\rho = \rho(x, t)$ is the density of the fluid at time t and position x. Now, the fluid mass within a volume V (given by the two spatial points x_1 and x_2, say $V = [x_1, x_2]$) at a time t, denoted by $M_{V,t}$, is given by the integral

$$M_{V,t} = \int_V \rho(x, t) \, dx = \int_{x_1}^{x_2} \rho(x, t) \, dx.$$

If we assume that mass cannot enter through the tube walls (i.e. there is no mass production or destruction), then mass within a given volume V can only change due to fluid flow. Let $v = v(x, t)$ be the velocity of the fluid at time t

and position x. The flow rate (or flux), denoted by $f = f(x,t)$ of fluid past a given position x at time t is given by

$$f(x,t) = \rho(x,t)v(x,t).$$

Thus, the rate of change in a volume V is

$$\frac{dM_{V,t}}{dt} = \frac{d}{dt}\int_{x_1}^{x_2} \rho(x,t)\ dx = \rho(x_2,t)v(x_2,t) - \rho(x_1,t)v(x_1,t).$$

Integration of both sides of the above equation over a time interval $[t_1, t_2]$ gives an expression of the mass evolution in terms of a pre-existing mass and an amount of mass flowing in and out of the volume at both boundaries:

$$\int_{x_1}^{x_2} \rho(x,t_2)\ dx = \int_{x_1}^{x_2} \rho(x,t_1)\ dx + \int_{t_1}^{t_2} \rho(x_2,t)v(x_2,t)\ dt - \int_{t_1}^{t_2} \rho(x_1,t)v(x_1,t)\ dt.$$

In order to derive the differential form of the conservation of mass, we assume that both ρ and v are differentiable. Then, with the relations

$$\rho(x,t_2) - \rho(x,t_1) = \int_{t_1}^{t_2} \frac{\partial\rho}{\partial t}(x,t)\ dt,\ \text{and}$$

$$\rho(x_2,t)v(x_2,t) - \rho(x_1,t)v(x_1,t) = \int_{x_1}^{x_2} \frac{\partial}{\partial x}\left(\rho(x,t)v(x,t)\right)\ dx,$$

we derive

$$\int_{t_1}^{t_2}\int_{x_1}^{x_2}\left[\frac{\partial}{\partial t}\rho(x,t) + \frac{\partial}{\partial x}\left(\rho(x,t)v(x,t)\right)\right]\ dxdt = 0.$$

Since this holds for arbitrary x_i and t_j ($i,j = 1,2$) the differential form of the conservation of mass reads

$$\partial_t\rho + \partial_x(\rho v) = 0. \tag{7.1}$$

For higher dimensions the conservation of mass can be derived in an analogous way yielding

$$\partial_t\rho + \nabla\cdot(\rho\mathbf{v}) = 0, \tag{7.2}$$

where $\mathbf{v}$ is a vector valued velocity now.

Note that with the transport theorem A.0.5, we could have derived (7.2) in a different way. Assuming that mass in a material volume is conserved over time, we can formulate the following equation:

$$\frac{d}{dt}\int_{V(t)} \rho\ dx = 0.$$

It states that the rate of change of mass in $V(t)$ is zero. Now, applying the transport theorem yields

$$\int_{V(t)} \left(\frac{\partial \rho}{\partial t} + \nabla \cdot (\rho \mathbf{v}) \right) dx = 0.$$

Since the above integral equation is true for any material volume $V(t)$, (7.2) follows.

Remark 7.1.1 *(Mass conservation in incompressible flow)*
An incompressible flow is defined by the property that the density of each material particle remains constant over time:

$$\rho(\mathbf{x}_{\mathbf{x}_0, t_0}(t), t) = \rho(\mathbf{x}_0, t_0).$$

In other words, $\frac{d\rho}{dt} = 0$. *From this and (7.2) it follows that* $\nabla \cdot \mathbf{v} = 0$, *since*

$$\nabla \cdot (\rho \mathbf{v}) = \rho \nabla \cdot \mathbf{v} + \mathbf{v} \cdot \nabla \rho.$$

It should be noted that this incompressibility definition does not contradict the usual definition by the statement of a solenoidal velocity field (i.e. $\nabla \cdot \mathbf{v} = 0$). In oceanography, this is often used as the definition of incompressibility and density is taken as an equation of state (e.g. $\rho = \rho(S, T, P)$, where S is salinity, T is temperature, and P is pressure, see sect. 7.1.4). Here, we claimed that density in each material particle is constant!

Another, often more useful alternative formulation of (7.2) in terms of the material derivative is given by

$$\frac{d\rho}{dt} + \rho \nabla \cdot \mathbf{v} = 0. \tag{7.3}$$

7.1.2 Conservation of Momentum

Newton's law of conservation of momentum ($\mathbf{v}m$, m the mass) states that the rate of change of momentum of an object is balanced by the sum of all forces on it. This can be generalized to fluids whose momentum is given as the integral of velocity times density in an (infinitesimal) material volume. In mathematical terms this relation is reflected in

$$\frac{d}{dt} \int_{V(t)} \rho \mathbf{v} \, dx = \int_{V(t)} \mathcal{F}_{\text{body}} \rho \, dx + \int_{S(t)} \mathcal{F}_{\text{surf}} \, ds. \tag{7.4}$$

The first integral on the right hand side of (7.4) represents the *internal (body) forces* in the material volume $V(t)$, while the second integral represents the *surface forces* on the surface $S(t)$ of the material volume.

We can apply the transport theorem A.0.5 to the left hand side of (7.4) which gives us:

$$\int_{V(t)} \left(\frac{\partial(\rho \mathbf{v})}{\partial t} + \nabla \cdot [(\rho \mathbf{v})\mathbf{v}] \right) dx = \int_{V(t)} \mathcal{F}_{\text{body}} \rho \, dx + \int_{S(t)} \mathcal{F}_{\text{surf}} \, ds.$$

Now, assuming that the surface forces can be represented by means of the *stress tensor* σ_{surf} by

$$\mathcal{F}_{surf} = \sigma_{surf}\mathbf{n},$$

where $\mathbf{n}$ is the outward unit normal on ds, and applying a scalar version of the divergence theorem A.0.1, we obtain

$$\int_{V(t)} \left(\frac{\partial(\rho\mathbf{v})}{\partial t} + \nabla \cdot [(\rho\mathbf{v})\mathbf{v}] \right) \, dx = \int_{V(t)} (\mathcal{F}_{body}\rho + \nabla\sigma_{surf}) \, dx.$$

Since this equation holds for arbitrary $V(t)$, we conclude with the differential form of the conservation of momentum:

$$\partial_t(\rho\mathbf{v}) + \nabla \cdot [(\rho\mathbf{v})\mathbf{v}] - \mathcal{F}_{body}\rho - \nabla\sigma_{surf} = 0. \tag{7.5}$$

In order to obtain a formulation with respect to the material derivative, we assume incompressibility and perform some more elementary transformations to obtain

$$\rho\frac{d\mathbf{v}}{dt} - \mathcal{F}_{body}\rho - \nabla\sigma_{surf} = 0. \tag{7.6}$$

Assuming that that we can restrict ourselves to the following three force terms

- the pressure gradient ∇p,
- the gravity acting on density as a potential ρg, and
- frictional forces summarized by a term $\mathcal{F} = \mathcal{F}(\mu, \mathbf{v})$,

where μ is the molecular viscosity, we obtain the momentum equation in analogy to [311]:

$$\rho\frac{d\mathbf{v}}{dt} + \nabla p - \rho g - \mathcal{F}(\mu, \mathbf{v}) = 0. \tag{7.7}$$

Note that often not a molecular viscosity is assumed but an eddy viscosity, assuming certain scale dependencies. Furthermore, we have assumed that the fluid is a *Newtonian fluid* (see e.g. [410] for a definition of Newtonian fluids; water and air can be considered Newtonian).

It should also be noted that equation (7.7) does not strictly state a conservation principle, since the force terms lead to (at least locally) diminishing momentum. Thus it is probably more correct to call it a "balance equation".

7.1.3 Conservation of Energy

Internal energy e, like density ρ or pressure p, is a state variable of a fluid flow system. The first law of thermodynamics states that the *total energy E* of a closed system is balanced by the the work applied to the system and the heat added to it. Thus, total energy is the sum of *kinetic energy*, potential energy, and internal energy. The kinetic energy per unit mass of a material particle is

$\frac{1}{2}\mathbf{v}^2$, where $\mathbf{v}$ is the velocity. In order to derive a conservation law for energy, we apply the first law of thermodynamics to a material volume $V(t)$:

$$\frac{d}{dt}\int_{V(t)} \rho E \ dx = W + Q, \tag{7.8}$$

where W is the rate of work acting on the fluid and Q is the rate of heat added. Body force works at a rate of $\mathbf{v} \cdot \mathcal{F}_{body}\rho \ dx$ and the surface force at a rate of $\mathbf{v} \cdot \mathcal{F}_{surf} \ ds$, which gives

$$W = \int_{V(t)} \mathbf{v} \cdot \mathcal{F}_{body}\rho \ dx + \int_{S(t)} \mathbf{v} \cdot \mathcal{F}_{surf} \ ds.$$

Using the same argument as in the derivation of momentum conservation in sect. 7.1.2, we obtain

$$W = \int_{V(t)} (\rho\mathbf{v} \cdot \mathcal{F}_{body} + \nabla(\mathbf{v} \cdot \sigma_{surf})) \ dx.$$

We assume a heating rate of q per unit of mass on each material particle and a heat flux of ς over the surface $S(t)$. Then

$$Q = \int_{V(t)} \rho q \ dx + \int_{S(t)} \varsigma \ ds.$$

Further on, we assume that heat diffusion is governed by Fourier's law, i.e.

$$\varsigma = k\mathbf{n} \cdot \nabla T,$$

where k is the *thermal conductivity* and T the temperature. Using this and the divergence theorem, we obtain

$$Q = \int_{V(t)} (\rho q + \nabla(k\nabla T)) \ dx.$$

Taking the above results, substituting into (7.8) and applying the transport theorem we get

$$\int_{V(t)} \left(\frac{\partial\rho E}{\partial t} + \nabla \cdot (\rho\mathbf{v}E)\right) \ dx = \int_{V(t)} (\rho\mathbf{v} \cdot \mathcal{F}_{body} + \rho q + \nabla(\mathbf{v} \cdot \sigma_{surf})$$
$$+ \nabla(k\nabla T)) \ dx.$$

Since this is true for arbitrary $V(t)$, we derived the *energy conservation equation*

$$\partial_t(\rho E) + \nabla \cdot [(\rho E)\mathbf{v}] = \rho\mathbf{v} \cdot \mathcal{F}_{body} + \rho q + \nabla(\mathbf{v} \cdot \sigma_{surf}) + \nabla(k\nabla T). \tag{7.9}$$

A more suitable form of the energy equation that can be derived by using the mass conservation equation (7.2) and some more elemental manipulations (see [410]). This yields an equation for the internal energy, which we give in a formulation involving a material derivative for e:

$$\rho \frac{de}{dt} = -p\nabla \cdot \mathbf{v} + \rho q + \nabla(k\nabla T) + \mathcal{H}. \tag{7.10}$$

The left hand side represents the advective energy transport which is balanced by the terms on the right hand side, which are

- the work done by pressure, $p\nabla \cdot \mathbf{v}$,
- external heating, ρq,
- diffusive energy transport, $\nabla(k\nabla T)$, and
- dissipative (frictional) heating $\mathcal{H}$, which is non-negative.

7.1.4 Equations of State – Closing the System

In the above sections, we derived a set of five equations:

1. conservation of mass (7.3), involving the density ρ and velocity $\mathbf{v}$;
2. conservation of momentum (7.7) for the three velocity components of $\mathbf{v} = (u, v, w)$, involving $\mathbf{v}$, ρ, and p;
3. conservation of energy (7.10), involving e, $\mathbf{v}$, T, ρ, and p.

Summarizing, we have seven state variables u, v, w, ρ, p, e, T but only five equations. Thus, in order to close the system, we need two more equations, the *equations of state*.

The perfect gas equation of state establishes a relation between pressure and density/temperature by

$$p = \rho R T, \tag{7.11}$$

where R is a constant specific to the individual gas. It is given by

$$R = \frac{R_g}{\mu m_u N_A},$$

where $R_g = 8.3143 \mathrm{JK}^{-1}\mathrm{mol}^{-1}$ is the gas constant, $N_A = 6.022 \cdot 10^{23}\ \mathrm{mol}^{-1}$ is Avogadro's number, $m_u = 1.66 \cdot 10^{-27}$ kg is the atomic mass unit and μ is the mean molecular weight. For dry air we have $R = 287 \mathrm{JK}^{-1}\mathrm{kg}^{-1}$.

For a perfect gas, the internal energy depends only on temperature, therefore we obtain a second equation of state

$$e = e(T). \tag{7.12}$$

The derivation of conservation laws and equations of state so far are valid for any fluid dynamics application in a non-rotating frame of reference. We have not specified any specialities for geophysical fluid dynamics or atmospheric fluid flow, and in particular, we did not consider the flow on a rotating sphere. We will postpone this to the appendix, and will proceed by simplifying the equations by certain assumptions.

If we assume that the relation of ρ and p can be described by a function ψ in the following sense:

$$\frac{1}{\rho}\delta p = \delta \psi,$$

where $\delta(\cdot)$ symbolizes a small variation in the corresponding function, then a flow is called *barotropic*. A common assumption is that $p = p(\rho)$ is a function of density alone, or that $\rho = c = $ const. in which case $\psi = c^{-1}p$.

Introducing specific entropy s, defined by $T\frac{ds}{dt} = \frac{de}{dt} + p\frac{d}{dt}\frac{1}{\rho}$ and assuming equations of state of the form

$$\rho = \rho(p, T), \quad s = s(p, T),$$

Pedlosky [311] derives several alternatives to the energy equation (7.10). For ideal gases the potential temperature equation is

$$\frac{d\Theta}{dt} = \frac{\Theta}{C_p T}\left(\frac{1}{\rho}\nabla(k\nabla T) + q\right),$$

with Θ the *potential temperature*, defined by $\Theta = T(\frac{p_0}{p})^{R/C_p}$, p_0 a reference pressure, and C_p the specific heat at constant pressure. Note that in the absence of internal and conductive heating (i.e. $q = 0$ and $\nabla(k\nabla T) = 0$), potential temperature is a conserved quantity for each material particle.

Another convenient form of the first thermodynamic law can be obtained from the assumption that ρ can be expressed as

$$\rho = \rho_0(1 - \alpha(T - T_0)).$$

Then the *heat equation* takes the form

$$\frac{dT}{dt} = \kappa \Delta T + \frac{q}{C_p},$$

where $\kappa = \frac{k}{\rho C_p}$ is the coefficient of thermal diffusivity.

7.1.5 Deriving the Shallow Water Equations

In global atmospheric and oceanic modeling, the principal assumption that the typical vertical length scale is small compared to the typical horizontal length scale is valid. A systematic scale analysis can be found in [311]. We will derive the shallow water equations along that line.

We consider the fluid with constant and uniform density, $\rho = $ const., and let the fluid be inviscid, which means that viscosity μ is negligible or in other words there are no internal frictional forces and $\mathcal{F}(\mu, \mathbf{v})$ vanishes. The height of the surface is given by $h = h(x, y, t)$, and the rigid bottom is given by $h_b = h_b(x, y)$. The body force from a potential ψ is modeled as a vector g

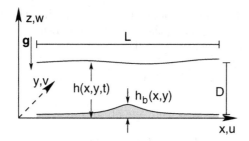

Fig. 7.1. Sketch of the shallow water set up

perpendicular to the horizontal (x, y)-plane, i.e. antiparallel to the z-axis. See fig. 7.1 for a sketch of the set up.

The specification of uniform and constant density leads to the incompressibility condition. This follows from mass conservation (7.2):

$$\nabla \cdot \mathbf{v} = 0.$$

Furthermore, the thermodynamics is decoupled from the system, leaving only the mass conservation and momentum equations.

Let D be the typical scale for the water depth (or the vertical length scale), while L be the typical horizontal length scale. Then the above principal assumption is given by

$$\delta = \frac{D}{L} \ll 1.$$

With the above assumptions and a rigorous scale analysis, not repeated here, one sees that the vertical velocity is at most of the order of $W \leq \mathcal{O}(\delta U)$, where W is the typical vertical and U the typical horizontal scale of velocity. Furthermore, the vertical pressure variation $\hat{p}$ can be neglected, where we wrote the total pressure $p = p(x, y, z, t) = -\rho g z + \hat{p}(x, y, z, t)$. This also yields the hydrostatic assumption (to order $\mathcal{O}(\delta^2)$ accuracy)

$$\frac{\partial p}{\partial z} = -\rho g.$$

Assuming a constant surface pressure p_0 one can write $p = \rho g(h - z) + p_0$. The independence of the horizontal components of the pressure gradient from z gives

$$\frac{\partial p}{\partial x} = \rho g \frac{\partial h}{\partial x}, \quad \frac{\partial p}{\partial y} = \rho g \frac{\partial h}{\partial y}.$$

One can reason that the horizontal velocities have to be independent of z. Therefore, the horizontal momentum equations (in component form) become

$$\partial_t u + u \partial_x u + v \partial_y u + g \partial_x h = 0, \tag{7.13}$$
$$\partial_t v + u \partial_x v + v \partial_y v + g \partial_y h = 0. \tag{7.14}$$

Using the z-independence of u and v, integrating the incompressibility condition above in z and taking the boundary conditions for w at the bottom $z = h_b$ and at the surface $z = h$ yields

$$\frac{\partial h}{\partial t} + \frac{\partial}{\partial x}\left[(h - h_b)u\right] + \frac{\partial}{\partial y}\left[(h - h_b)v\right] = 0.$$

Defining a total depth $H = h - h_b$ one obtains

$$\partial_t H + \partial_x(Hu) + \partial_y(Hv) = 0, \tag{7.15}$$

which together with (7.13) and (7.14) form the *shallow water equations*.

Using the material derivative and using vector notation $\xi_H = (\xi|_x, \xi|_y)$, with $\xi|_x$ and $\xi|_y$ the x and y components of ξ_H resp., for the horizontal components, one can rewrite the system to the following form

$$\frac{d\mathbf{v}_H}{dt} + g\nabla_H h_H = 0, \tag{7.16}$$

$$\frac{dH}{dt} + H\nabla_H \cdot \mathbf{v}_H = 0. \tag{7.17}$$

In global atmospheric modeling, the rotating shallow water equations in spherical geometry are often used for testing numerical schemes. A standard set of test cases and a collection of formulations of shallow water equations can be found in [415]. We list some additional forms in the appendix C. We did not report on the full 3-dimensional set of equations that are usually used in numerical models of the atmosphere. We refer to the standard literature (e.g. [311]). Some 3D adaptive model formulations can be found in [20, 221, 308].

7.2 Finite Volume Methods

We are now going to explore methods for discretizing the equations of fluid motion, derived so far. One of the objectives of our discretization is to preserve structure of the equations under discretization. For instance, if we discretize the continuity equation (or conservation of mass equation) then the discrete form shall also conserve mass.

One numerical scheme that is especially suited for discretization of conservation laws is the finite volume method. We will start the presentation with a one-dimensional example.

7.2.1 One-Dimensional Model Problem and Basic Algorithmic Idea

In order to derive the basic idea behind finite volume methods (FVM), we start with a 1D model problem, namely the advection equation in flux form for a density distribution $\rho = \rho(x,t)$:

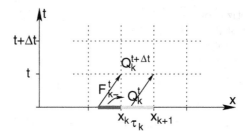

Fig. 7.2. Idea of the finite volume method: F_{k-}^t represents the flux from the upstream side of cell τ_k, contributing to the mean value $Q_k^{t+\Delta t}$. The figure shows Lagrangian fluxes that are translated into the Eulerian flux contribution for each cell interface in finite volume methods

$$\partial_t \rho + \partial_x F(\rho) = 0, \tag{7.18}$$

where $F = u\rho$ is the flux function. By using the transport theorem A.0.5 we get an integral form of (7.18)

$$\frac{d}{dt} \int_{\mathcal{G}} \rho(x,t) \, dx = 0, \tag{7.19}$$

where $\mathcal{G}$ is the whole domain. Assume we have a fixed mesh of cells τ_i, $i = 1 : N$, then the mass in volume τ_k can only change by inflow and outflow over the left and right borders of τ_k (see fig. 7.2). This can be formulated by integrating (7.18) over the time interval $[t, t + \Delta t]$ and the grid cell $\tau_k = [x_k, x_k + 1]$:

$$\int_{\tau_k} \rho(x, t+\Delta t) \, dx = \int_{\tau_k} \rho(x,t) \, dx + \int_t^{t+\Delta t} F(\rho(x_{k+1}, t)) \, dt - \int_t^{t+\Delta t} F(\rho(x_k, t)) \, dt. \tag{7.20}$$

Taking Q_k^t as a discrete mean value of ρ in the grid cell τ_k, i.e. $Q_k^t \approx \frac{1}{\Delta x} \int_{\tau_k} \rho(x,t) \, dx$, where we assumed that $|\tau_k| = x_{k+1} - x_k = \Delta x$, and defining the time averaged mass fluxes

$$F_{k-} \approx \frac{1}{\Delta t} \int_t^{t+\Delta t} F(\rho(x_k, t)) \, dt, \quad \text{and}$$

$$F_{k+} \approx \frac{1}{\Delta t} \int_t^{t+\Delta t} F(\rho(x_{k+1}, t)) \, dt,$$

we obtain the discrete finite volume representation of (7.20)

$$Q_k^{t+\Delta t} = Q_k^t + \frac{\Delta x}{\Delta t} \left(F_{k+} - F_{k-} \right). \tag{7.21}$$

This equation represents the main idea of finite volume methods. Now, there are a lot of more or less subtle issues in order to construct a stable, efficient and accurate discrete method from this principle. There are choices to be taken for

- the time discretization $\frac{d}{dt}$ in (7.19);
- the calculation of fluxes $F_{k\pm}$;
- the cell averages Q_k^t;
- the finite volumes τ_k.

For these details the reader is referred to [263, 264].

7.2.2 Multidimensional Extension

A multi-dimensional representation of (7.18) reads

$$\partial_t \rho + \nabla \cdot F(\rho) = 0, \qquad (7.22)$$

where $F(\rho) = \rho \mathbf{v}$ is the flux function, and $\mathbf{v}$ is a given velocity.

An equivalent form of this conservation law is given by the integral form

$$\frac{d}{dt} \int_V \rho(\mathbf{x}, t) \, dx = 0,$$

where we have used the transport theorem A.0.5. Here, V denotes the whole computational domain as a control volume. Furthermore, we assumed that no boundary conditions impose a flux of material into or out of the domain.

In order to derive a finite volume scheme, we consider a (small) reference volume V. Since the above conservation law (7.22) represents conservation of mass, we state that mass within V can only change due to mass fluxes over the boundary ∂V:

$$\frac{d}{dt} \int_V \rho(\mathbf{x}, t) \, dx = - \oint_{\partial V} f(\xi, t) \cdot \mathbf{n}(\xi) \, dx, \qquad (7.23)$$

where $\xi \in \partial V$ is on the boundary of V, $\mathbf{n}$ is the outer normal unit vector, and f represents the contribution over ∂V. Note that by the minus sign, we state that f denotes the inward contribution. Integrating both sides over the time interval of one time step $I = [t, t + \Delta t]$ yields:

$$\int_V \rho(\mathbf{x}, t + \Delta t) \, dx = \int_V \rho(\mathbf{x}, t) \, dx + \int_t^{t+\Delta t} \oint_{\partial V} f(\xi, t) \cdot \mathbf{n}(\xi) \, dx dt. \quad (7.24)$$

Note that we assume V to be fixed (but arbitrary) in time and space.

Let us assume that there is a mesh with simplicial cells (for triangular and quadrilateral cells see chap. 3). Let the reference volume $V = \tau$ be a cell with edges/faces e_j, $j = 1 : M$.

Algorithm 7.2.1 *(Basic finite volume method)*

A finite volume method (FVM) can be derived utilizing the following components. For each cell $\tau \in \mathcal{T}$ in the triangulation:

1. Let Q_τ^t be the cell mean value of ρ,

$$Q_\tau^t \approx \frac{1}{|\tau|} \int_\tau \rho(\mathbf{x}, t) \; dx,$$

where $|\tau|$ is the area/volume of cell τ.
2. Approximate the in-/outflow over cell edges/faces by

$$F_j^t \approx \frac{1}{\Delta t} \oint_t^{t+\Delta t} \left(\frac{1}{h_j} \int_{e_j} f(\xi, t) \cdot \mathbf{n} \; dx \right) \; dt,$$

where h_j denotes the edge length or face area resp., and F_j^t is the time averaged mean edge/face flux.
3. Then equation (7.24) can be discretized by

$$Q_\tau^{t+\Delta t} = Q_\tau^t + \frac{\Delta t}{|\tau|} \sum_{j=1:M} h_j F_j^t. \tag{7.25}$$

Note that we have to divide by $|\tau|$ and multiply by h_j and Δt in order to reproduce (7.24). Furthermore, the terms on the right hand side only depend on time t.

There is a large family of different finite volume methods. The main differences concern the computation of fluxes F_j^t (see [264] for a comprehensive presentation of finite volume methods, and Thuburn for shape preserving flux computation [392]).

Note that a solution of the above algorithm may be discontinuous. The algorithm requires a solution to be piecewise continuous only. Therefore, FVM are well suited for problems that develop shocks. A typical class of equations that are solved with FVM are hyperbolic conservation laws.

In the simplest case, where Q_τ represents a constant value in the cell τ, in general we will have that $Q_{\tau_j} \neq Q_{\tau_i}$ for two distinct but neighboring cells $\tau_i \cap \tau_j = e_j$. Then calculating the flux over e_j corresponds to solving a one-dimensional *Riemann problem* in the normal direction with respect to e_j. Methods that utilize Riemann problems to derive numerical flux functions F_j are called *Godunov's methods*. Examples for Godunov's method in an adaptive mesh refinement atmospheric simulation can be found in [207] and in combination with mesh moving adaptivity in [13]. On triangular uniform meshes Pesh has implemented a finite volume scheme for the shallow water equations in the scalar potential vorticity-divergence formulation [316]. Choi and coworkers present a spectral finite volume method for the shallow water equations [85]. An upwinding finite volume type method for the shallow water equations can be found in [160] while a comparison of several upwind difference methods is given in [172]. Botta et al. propose a finite volume scheme that is well balanced in the sense that it is numerically stable for an ill conditioned pressure gradient in terrain following coordinates [67]. Another approach to a similar gravity wave related problem is given in [242], utilizing a finite volume method with an orography adapted grid. A general approach of finite volume methods on unstructured meshes can be found in [115], while a high-order accurate scheme for unstructured meshes is given in [305]. Generally constraint-preserving finite volume schemes for advection are given in [395].

7.2.3 Convergence, Consistency and Stability

In order to assess the convergence of the scheme, we need to introduce some notation. We will use the fundamental theorem of numerical analysis which is well known as *Dahlquist's equivalence theorem* for ordinary differential equations or as the *Lax equivalence theorem* for partial differential equations. It states that a method converges if and only if it is consistent and stable. Let the numerical method be given by the evolution operator Ψ that maps a solution at time t to a new solution at time $t + \Delta t$:

$$\Psi : \mathcal{V} \to \mathcal{V}$$
$$Q^t \mapsto Q^{t+\Delta t} = \Psi(Q^t),$$

where $\mathcal{V}$ is the solution space, and Q^t is a discrete solution (e.g. the cell mean values for all cells) at time t. Then we can define the one-step error:

Definition 7.2.2 *(One-step error, truncation error)*
   The one-step error ε^1 *is defined by*

$$\varepsilon^1 = \Psi(q^t) - q^{t+\Delta t},$$

where q^t is the true solution at time t. The local truncation error ε^t is defined as the one-step error scaled by the time step:

$$\varepsilon^t = \frac{1}{\Delta t} \left(\Psi(q^t) - q^{t+\Delta t} \right).$$

A simple possibility to compute the mass flux between cells consists of the first order finite difference scheme $F_j = \mathbf{n}_j \cdot \mathbf{v}(Q_{T_j} - Q_{T_i})$. Here $\mathbf{n}_j$ denotes the outward unit normal vector on edge e_j. Calculating the numerical flux in this way, results in a first order upwind method. One can show that the truncation error for the upwind scheme is of order

$$\varepsilon^t = \mathcal{O}(\Delta x) + \mathcal{O}(\Delta t^2),$$

which means that it is dominated by the first order spatial error [264]. We say that the upwind scheme is of *first order consistency*. Essentially, this means that when we decrease the mesh size by a factor of two, then the local truncation error will also decrease at the same rate.

   Now, that we showed consistency, we have to tackle the stability of the method. The stability assessment is concerned with a global bound of the error when the time step approaches zero ($\Delta t \to 0$), and hence the number n of time steps approaches infinity ($n \to \infty$). To see the principle, we assume that in the n-th time step $t = n\Delta t$ there is a corresponding error ε^n such that

$$Q^t = q^t + \varepsilon^n.$$

Using the evolution representation, we obtain $Q^{t+\Delta t} = \Psi(Q^t) = \Psi(q^t + \varepsilon^n)$. From this we deduce for the global error:

$$\varepsilon^{n+1} = Q^{t+\Delta t} - q^{t+\Delta t} = \left\{\Psi(q^t + \varepsilon^n) - \Psi(q^t)\right\} + \left\{\Psi(q^t) - q^{t+\Delta t}\right\} \quad (7.26)$$
$$= \left\{\Psi(q^t + \varepsilon^n) - \Psi(q^t)\right\} + \Delta t \varepsilon^t.$$

Thus, the global error is defined by two terms: the one-step error of the current time step, and the error induced by the numerical method (represented by the evolution operator) applied to the perturbed data at time t. This second term is bounded in a stable numerical scheme.

If Ψ is a contractive operator, i.e.

$$\|\Psi(Q_1) - \Psi(Q_2)\| \le \|Q_1 - Q_2\|,$$

for some norm $\|\cdot\|$ and two discrete functions Q_1 and Q_2, then from (7.26) we get

$$\|\varepsilon^{n+1}\| \le \|\Psi(q^t + \varepsilon^n) - \Psi(q^t)\| + \Delta t \|\varepsilon^t\|$$
$$\le \|\varepsilon^n\| + \Delta t \|\varepsilon^t\|.$$

Applying this recursively for all n time steps and taking $\|\varepsilon_{\max}\| = \max_{0 \le t \le n\Delta t} \|\varepsilon^t\|$, we get

$$\|\varepsilon^n\| \le \|\varepsilon^0\| + t \|\varepsilon_{\max}\|, \quad \text{with } t = n\Delta t.$$

$\|\varepsilon^0\|$ is the error in the initial data (continuous versus gridded data) and we require $\|\varepsilon^0\| \to 0$ for $\Delta t \to 0$ in order to solve the correct initial value problem. If the method is consistent (see above), then also $\|\varepsilon_{\max}\| \to 0$ for $\Delta t \to 0$, which proves the stability and therefore the convergence of the method.

Note that a relaxed condition on the operator Ψ is sufficient to prove convergence:

$$\|\Psi(Q_1) - \Psi(Q_2)\| \le (1 + \alpha \Delta t) \|Q_1 - Q_2\|,$$

where α is a constant independent of Δt. If the method is linear, i.e. Ψ is a linear operator, then the analysis becomes even more simple, since $\Psi(q^t + \varepsilon^n) = \Psi(q^t) + \Psi(\varepsilon^n)$. So from the requirement

$$\|\Psi(q^t + \varepsilon^n) - \Psi(q^t)\| = \|\Psi(\varepsilon^n)\| \le (1 + \alpha \Delta t) \|\varepsilon^n\|$$

we deduce that $\|\Psi\| \le 1 + \alpha \Delta t$. It suffices for Ψ to be bounded by a constant ($\|\Psi\| \le C$) to show stability of a linear method. This is referred to as the *Lax-Richtmyer stability*. One can show that the above given upstream method is a linear method fulfilling the requirement of a bounded evolution operator.

A *von Neumann stability analysis* for the upwind method discretizing the advection equation

$$Q_T^{t+\Delta t} = (1 - \nu)Q_T^t + \nu Q_{T_{\text{upwind}}}^t$$

yields that the method is stable if the *Courant number* $\nu = \mathbf{v}\frac{\Delta t}{\Delta x}$ fulfils $0 \leq \nu \leq 1$. This stability criterion is referred to as the *Courant Friedrichs Levy (CFL) stability condition* [102].

Generally, FVM are restricted by the CFL condition, which makes them less advantageous for adaptive methods, since the time step is restricted by the smallest grid size. However, many current AMR methods rely on FVM, and employ time sub stepping in order to retain stability while maintaining efficiency even with locally refined grids [14, 52, 54, 55, 283, 291, 384]. A more rigorous convergence analysis can be found in [94] or [301]. Stability of time-implicit finite volume schemes has been studied in [158]. Extensions to the finite volume methods and to the computation of fluxes on unstructured tetrahedral meshes can be found in [11]. A relaxation of the CFL condition has been proposed in a so called h-box method [53]. A convergence proof and relation of higher order finite volumes to discontinuous Galerkin methods is given in [246]. Spectral finite volumes are employed in [407] to solve conservation laws in 2D.

Lin and Rood [269, 268] extend the finite volume approach to long time steps by observing that fluxes can be computed by following the Lagrangian trajectories of the cells. They use a splitting scheme for the 2D case, where F_x and F_y are the 1D flux functions in x- and y-direction respectively. Then the splitting scheme can be written as

$$Q^{t+\Delta t} = Q^t + F_x\left[Q^t + \frac{1}{2}F_y(Q^t)\right] + F_y\left[Q^t + \frac{1}{2}F_x(Q^t)\right].$$

Now the fluxes are decomposed into an integer part (corresponding exactly to integer multiples of the time step) and a fractional part that can be computed by any available Eulerian upstream flux form scheme. This method is shown to be unconditionally stable for uniform flow fields. The original CFL stability criterion is replaced by a so called *Lipschitz stability criterion* that states that the trajectories must not cross or that the upstream cell must not collapse (see also [371]). Jablonowsky uses the Lin-Rood scheme in an adaptive dynamical core [221]. Lanser and coworkers proposed an Osher type finite volume scheme for shallow water equations on the sphere [253, 254].

7.3 Discontinuous Galerkin Methods

Discontinuous Galerkin methods have been introduced by Reed and Hill [334] and a good overview of the state of the art up to 1999 can be found in [95]. A comparison of different finite difference, finite volume and discontinuous

Galerkin (DG) methods has been conducted by Shu [365]. An extension to nonlinear hyperbolic conservation laws can be found in [92]. Today, discontinuous Galerkin methods have been applied to all kinds of problems, including elliptic partial differential equations [12]. An adaptive DG method based on an orthogonal basis is proposed in [336]. The popularity of DG methods stems from their scalability in parallel applications and their accuracy when combined with high order approximation methods.

7.3.1 The Basic Idea for Ordinary Differential Equations

In order to derive the discontinuous Galerkin method, we start with the one-dimensional case again. Assume, we want to solve an ordinary differential equation (ODE) of the form

$$\frac{d\rho(t)}{dt} = f(t)\rho(t), \tag{7.27}$$

on the time interval $t \in I = [0, T]$ and with initial condition $\rho(t = 0) = \rho_0$. We will denote the discrete approximation to the true solution ρ of (7.27) by ρ_h. For the discontinuous Galerkin (DG) discretization, the interval I is partitioned into sub-intervals $I^n = [t^n, t^{n+1}]$, $n = 0 : N - 1$, where $t^0 = 0 < \cdots < T = t^N$. The discrete solution ρ_h is determined on each $I^{(n)}$ as the polynomial of degree at most $p^{(n)}$ (the polynomial degree is allowed to change on the sub-intervals) such that

$$-\int_{I^{(n)}} \rho_h(s)\frac{db(s)}{dt}\, ds + \hat{\rho}_h b\big|_{t^n}^{t^{n+1}} = \int_{I^{(n)}} f(s)\rho_h(s)b(s)\, ds, \tag{7.28}$$

for all polynomials b of at most degree $p^{(n)}$. This is the classical weak formulation after using integration by parts on the left hand side to shift over the derivative from the solution to the test function b. Note that the link between two adjacent intervals is only due to the trace function $\hat{\rho}_h$. In the simple case of constant polynomial degree $\hat{\rho}_h$ can be interpreted as a flux function and the DG method is a representative of the finite volume methods. More precisely

$$\hat{\rho}_h(t^n) = \begin{cases} \rho_0, & \text{if } t^n = 0, \\ \lim_{\epsilon \downarrow 0} \rho_h(t^n - \epsilon), & \text{otherwise.} \end{cases}$$

Here $\lim_{\epsilon \downarrow 0} \rho_h(t^n - \epsilon)$ denotes the upstream value of ρ_h at the interface t^n.

We summarize the main properties of DG methods and refer for more details to the literature (e.g. [93, 95, 96]).

- DG methods comprise discontinuous discrete solution functions ρ_h, so that no inter-element continuities have to be considered. This feature makes DG methods well suited for hp-adaptive methods and also efficient in parallel computing environments.

- DG methods utilize a weak formulation of the problem, which makes them preferable for problems with irregular solution behavior.
- DG methods rely on the definition of a suitable interface function, the numerical trace function $\hat{\rho}_h$. The trace function is crucial to the convergence of DG methods.
- The method introduced so far is consistent, if for the true solution $\hat{\rho} = \rho$ holds.
- For suitable choices of the numerical trace function $\hat{\rho}_h$ one can prove the stability of the method.
- DG methods are locally conservative.

7.3.2 Application of DG Methods to Multi-Dimensional Scalar Conservation Laws

We will first apply the DG method to the spatial discretization and later extend the DG method by Runge-Kutta time discretization. Let us consider the d-dimensional conservation law (7.22) again

$$\rho_t + \nabla \cdot F(\rho) = 0,$$

with $F(\rho) = \mathbf{v}\rho$, ρ and $\mathbf{v}$ functions of $(\mathbf{x}, t)$, $\mathbf{x} \in \mathbb{R}^d$, $t \in [0, T]$ and an initial condition $\rho|_{t=0} = \rho_0$.

The weak form of the above equation with a discrete (piecewise polynomial) function ρ_h is given by

$$\int_\tau (\rho_h)_t b \; dx - \int_\tau \mathbf{v}\rho_h \cdot \nabla b \; dx + \int_{\partial\tau} \widehat{\mathbf{v}\rho_h} \cdot \mathbf{n} b \; ds = 0, \qquad (7.29)$$

where b denotes a test function, τ denotes an element of a triangulation $\mathcal{T}$ of the computational domain $\mathcal{G} \subset \mathbb{R}^d$, and $\mathbf{n}$ is the outward normal unit vector with respect to τ. If we choose the numerical trace function to be

$$\widehat{\mathbf{v}\rho_h} = \mathbf{v} \lim_{\epsilon \downarrow 0} \rho_h(\mathbf{x} - \epsilon\mathbf{v}, \cdot),$$

then the corresponding DG method coincides with the classical upwind finite volume scheme. Again, this method is locally conservative, i.e.

$$\frac{d}{dt} \int_\tau \rho_h \; dx + \int_{\partial\tau} \widehat{\mathbf{v}\rho_h} \cdot \mathbf{n} \; ds = 0.$$

This follows from setting $b \equiv 1$ in (7.29). Note that τ is a fixed volume, so we have to consider the change of mass due to fluxes over the cell's boundary $\partial\tau$.

The above space discretization of the scalar conservation law can be combined with a Runge-Kutta (RK) scheme for the time discretization to form the *Runge-Kutta Discontinuous Galerkin* (RKDG) methods. Using a multi-step explicit RK scheme allows very efficient and scalable parallel solution of the conservation equation. Starting with (7.29), we reformulate the equation substituting $F(\rho) = \rho\mathbf{v}$ again

$$\int_\tau (\rho_h)_t b \ dx - \int_\tau F(\rho_h) \cdot \nabla b \ dx + \int_{\partial\tau} \hat{F}(\rho_h) \cdot \mathbf{n}b \ ds = 0.$$

Discretizing the integrals we see easily, that we obtain a system of ordinary differential equations of the form

$$\frac{d\rho_h}{dt} = L(\rho_h),$$

where L is a linear operator. This ODE can be discretized by a multi-step RK scheme in the following way:

Algorithm 7.3.1 *(Multi-step Runge-Kutta DG method)*

1. *Set $\rho_h^{(0)} := \rho_h^t$, where ρ_h^t is the solution of the conservation law at time t.*
2. *For $k = 1 : K$ compute intermediate functions*

$$\rho_h^{(k)} = \sum_{j=0:k-1} \alpha_{k,j} w_h^{k,j}, \quad w_h^{k,j} = \rho_h^{(j)} + \frac{\beta_{k,j}}{\alpha_{k,j}} \Delta t L_h \left(\rho_h^{(j)} \right),$$

   with L_h the discretized form of the operator L above, and $\alpha_{k,j}$, $\beta_{k,j}$ given by the corresponding RK scheme.
3. *Set $\rho_h^{t+\Delta t} = \rho_h^{(K)}$.*

Note that, in order to obtain a stable method, there are more requirements on the RK scheme. Usually, the RKDG method has to be combined with a generalized slope limiter that stabilizes the forward Euler step $\rho_h^{(k)} \to w_h^{k,j}$. For the details consult [28] and the references therein. Another stability constraint follows from the von Neumann analysis of the RK method. When combining a $(k+1)$-step RK method with a DG method with local polynomials of at most degree k, then we obtain an method of order $(k+1)$ accuracy, provided that

$$|\mathbf{v}|\frac{\Delta t}{\Delta x} \leq \frac{1}{2k+1},$$

which, for high orders of approximation, is a severe CFL restriction.

RKDG methods are used by Giraldo and coauthors in an application to the spherical shallow water equations [169]. A spectral element based DG method for the shallow water equations has been proposed in [140]. Nair and coauthors use a total variation diminishing RK scheme for the time integration that renders additional filtering of high frequency oscillations obsolete [297, 296][1]. Discontinuous Galerkin methods are suited for multi-scale problems that comprise hyperbolic and parabolic character [81]. DG methods can also

---

[1] It should be noted that TVD schemes have to be applied in time and space to yield non-oscillatory filter-free RKDG methods.

be combined with ADER schemes, known to yield high order in finite volume discretizations [132]. An implicit-explicit scheme with DG approach can be found in [211]. Iskandarani et al. compared several schemes, including DG and finite volume schemes for advection equations in [216]. A convergence analysis of DG methods for advection dominated flows in the case of discontinuous solutions is given in [402] while smooth solutions are considered in [427].

7.4 Conservative Semi-Lagrangian Methods

Semi-Lagrangian schemes, probably introduced in the late 1950's by Wiin-Nielsen [412] (see also [343, 351]), have gathered a wide acceptance for solving advection-dominated problems, especially in the atmospheric sciences. An overview of the state of the art of semi-Lagrangian methods up to 1990 can be found in [374]. Combining the semi-Lagrangian characteristics based time discretization with a finite element space discretization yields the Lagrange-Galerkin methods [56, 129, 190, 380]. These methods are appreciated for their

- numerical stability, allowing long time-steps,
- accuracy, which allows arbitrary high order approximation, and
- algorithmic simplicity facilitating the parallelization and easy implementation.

One of the main drawbacks of the method, however, is the lack of conservation properties in its original formulation. Therefore, several investigations aimed to find semi-Lagrangian procedures that ensure conservation of mass or other quantities. Some of the early approaches just fixed the global mass in the advection procedure more or less intelligently (see e.g. [57, 327, 381]). Other approaches rely on integrating the upstream concentration function and preserve the integral along the trajectory, like in [329]. In [298] the integral form is used together with a constrained interpolation approach to gain a conservative semi-Lagrangian method.

A conceptually different approach is to use geometric properties of the flow. Here, volumes are advected along trajectories and physical values are preserved within these volumes. For rectangular and uniform meshes Scroggs and Semmazzi [360] and Machenhauer and Olk [271] have shown geometry based approaches. Priestley applied this to triangular meshes [328]. Peng et al. have implemented a conservative semi-Lagrangian scheme in the framework of MM5, the meso-scale model of Pennsylvania State University and NCAR [313].

The Eulerian-Lagrangian Localized Adjoint Method (ELLAM) has been introduced by Celia et al. [80] and Russell and Trujillo [349] (see also [406]). An arbitrary Lagrangian-Eulerian method suitable for moving grids is proposed in [162]. An efficient remapping scheme for Eulerian-Lagrangian schemes for triangular meshes can be found in [363], and an ELLAM scheme for unstructured meshes with non-uniform time stepping is proposed in [423]. These methods are related to the weak Lagrange-Galerkin method used by Giraldo [168] and introduced by Benque and Labadie [249]. For the characteristic Galerkin method, see [267]. Essentially, the weak form of the conservation form of the equation is integrated at upstream positions. The difficulty consists of transferring the upstream integrated values to downstream grid points. Recently, Chen proposed a combination of characteristics based (semi-Lagrangian) and discontinuous (nonconforming) finite element methods [83]. A combination of Eulerian-Lagrangian and finite volume methods that circumvents the CFL restriction for finite volume schemes can be found in [192]. A forward-in-time semi-Lagrangian conservative method can be found in [377]. Another short characteristics based approach has been proposed in [112]. Falcone and Ferretti apply a semi-Lagrangian method to evolutive Hamilton-Jacobi equations and prove certain conservation properties. For 1D implementations they even show a correspondence to Godunov methods [143]. Israeli and coworkers combine a semi-Lagrangian scheme with a conservative splitting scheme to obtain an unconditionally stable and conservative method for the shallow water equations [219]. Finally, we like to mention the possibility of even extending the framework of Lagrangian techniques to the broad field of geometric integrators (see [265]).

Although some of the above papers mention the possibility of extending the respective approach to unstructured, adaptively refined meshes, there are hardly any demonstrations of these assumptions. Recently, Iske and Käser presented work related to the schemes presented here [218]. Some of the material in this chapter has been published in [42]. All these methods are suitable for adaptive mesh refinement.

In order to set up a framework for conservative semi-Lagrangian advection, let us first consider the (Eulerian) flux form formulation of conservation of a density, say $\rho = \rho(\mathbf{x}, t) \in \bar{\mathcal{G}}$:

$$\frac{\partial \rho}{\partial t} + \nabla \cdot (\rho \mathbf{v}) = 0 \quad \text{in } \bar{\mathcal{G}}, \tag{7.30}$$

where the computational domain is defined by $(\mathbf{x}, t) \in \bar{\mathcal{G}} = \mathcal{G} \times I$, $\mathcal{G} \subset \mathbb{R}^d$, $d = 2, 3$, the spatial domain, $I \subset [0, \infty[$ the time interval, and $\mathbf{v} = \mathbf{v}(\mathbf{x}, t) \in \mathbb{R}^d$ is a given multi-dimensional flow. On the other hand, the Lagrangian formulation of the homogeneous advection equation, assuming a divergence-free wind $\mathbf{v}$, is given by:

$$\frac{d\rho}{dt} = 0 \quad \text{on } \bar{\mathcal{G}}, \tag{7.31}$$

where $\frac{d\rho}{dt} = \frac{\partial \rho}{\partial t} + \mathbf{v} \cdot \nabla \rho$ is the material derivative. For being well defined, (7.30) and (7.31) need boundary conditions $\rho(\mathbf{x}_b, t) = \rho_b(\mathbf{x}_b, t) \forall \mathbf{x}_b \in \partial \mathcal{G}$, and initial conditions $\rho(\mathbf{x}, t = 0) = \rho_0(\mathbf{x}) \forall \mathbf{x} \in \mathcal{G}$, suitably chosen. In the following, we will not consider boundary conditions, so we assume that $\mathcal{G} = \mathbb{R}^d$.

Equation (7.31) in this form is not conservative, so we have to consider a source term with the divergence operator to obtain an equivalent to (7.30):

$$\frac{d\rho}{dt} = \rho \nabla \cdot \mathbf{v} \quad \text{on } \bar{\mathcal{G}}. \tag{7.32}$$

In order to construct mass conserving semi-Lagrangian advection schemes, we want to consider the integral (conservation) form of the advection equation that can be obtained for example by the transport theorem A.0.5:

$$\frac{d}{dt} \int_{V(t)} \rho \, dx = 0, \tag{7.33}$$

with $V(t)$ a reference volume moving with the flow. (7.33) will be discretized by the semi-Lagrangian method in the following section. We will derive definitions for local and global mass conservation from that.

The proposed algorithms either discretize the integrals by numerical quadrature schemes or by geometric intersection. Another approach, namely the MPSLM scheme, discretizes the integrals by a high finite number of mass cells, which essentially form a subdivision of grid cells.

7.4.1 Adaptive Semi-Lagrangian Scheme

In this section we give a brief introduction to the adaptive semi-Lagrangian scheme, introduced in [34]. We first consider equation (7.31) and use a time-centered discretization given by

$$\frac{\rho(\mathbf{x}, t + \Delta t) - \rho(\mathbf{x} - 2\alpha(\mathbf{x}), t - \Delta t)}{2\Delta t} = 0 \quad \text{in } \bar{\mathcal{G}}, \tag{7.34}$$

where Δt is the discrete time step, $\alpha(\mathbf{x})$ denotes the path to the upstream position corresponding to $\mathbf{x}$. $\alpha(\mathbf{x})$ is given by a simple ordinary differential equation, namely

$$\dot{\mathbf{x}} = \frac{d\mathbf{x}}{dt} = \mathbf{v}(\mathbf{x}, t), \tag{7.35}$$

where $\mathbf{v} = \mathbf{v}(\mathbf{x}, t)$ is the given velocity field as in (7.30). Solving (7.35) for $t + \Delta t$ and taking the difference of the positions gives $\mathbf{x}(t + \Delta t) - \mathbf{x}(t - \Delta t) = 2\alpha(\mathbf{x})$.

We end up with three steps of the so called *three-time-level algorithm*[2] :

Algorithm 7.4.1 *(Basic semi-Lagrangian Algorithm)*

1. *Calculate $\alpha(\mathbf{x})$ by solving (7.35).*
2. *Interpolate upstream value $\rho(\mathbf{x} - 2\alpha(\mathbf{x}), t - \Delta t)$.*
3. *Update grid value $\rho(\mathbf{x}, t + \Delta t)$ using (7.34).*

The given algorithm can be modified into a two-time-level algorithm by observing the fact that two decoupled calculations occur, when stepping forward in time: one acting on even, the other one on odd time steps. (7.34) is then modified to

---

[2] The algorithm is called *three-time-level algorithm*, because it spans over three time levels, $t + \Delta t$, t, and $t - \Delta t$

$$\frac{\rho(\mathbf{x}, t) - \rho(\mathbf{x} - \alpha(\mathbf{x}), t - \Delta t)}{\Delta t} = 0 \implies \rho(\mathbf{x}, t) = \rho(\mathbf{x} - \alpha(\mathbf{x}), t - \Delta t), \quad (7.36)$$

where we assume time level t to be unknown. Discretizing the ODE by a fixed point iteration, we obtain the discrete form of (7.35):

$$\alpha^{k+1}(\mathbf{x}) = \Delta t \cdot \mathbf{v}\left(\mathbf{x} - \frac{\alpha^k(\mathbf{x})}{2}, t + \frac{\Delta t}{2}\right). \quad (7.37)$$

There are many more possibilities to solve the ODE (7.35). A comprehensive comparison for this part of the computation can be found e.g. in [7]. In [281] another way of improving the computation of α is demonstrated, while a semi analytical way to determine departure points is proposed in [276]. The convergence has been studied by McDonald [279, 280] and Falcone and Ferretti [142]. An analysis for the related Lagrange-Galerkin methods for advection-dominated flows can be found in [31]. Another study of the stability is presented in [170].

So far, we have not mentioned a mesh. In fact, this algorithm in principal works on arbitrary meshes and even in mesh-less situations [41]. However, we consider adaptively refined triangular and tetrahedral meshes in 2D and 3D here.

In order to formulate the adaptive semi-Lagrangian algorithm for equation (7.36), two different meshes have to be considered, one at time t which will be modified according to a suitable refinement criterion, and one fixed mesh at time $t - \Delta t$. We denote the k-th iterate of the new mesh with $\mathcal{T}^{(k)}(t)$ and the old fixed mesh with $\mathcal{T}(t - \Delta t)$. We embed the semi-Lagrangian algorithm into an adaptive iteration to obtain:

Algorithm 7.4.2 *(Adaptive semi-Lagrangian Algorithm)*

1. *Duplicate mesh $\mathcal{T}(t - \Delta t)$, obtaining an initial mesh for the adaptive iteration $\mathcal{T}^{(0)}(t)$.*
2. WHILE $\mathcal{T}^{(k)}(t) \neq \mathcal{T}^{(k-1)}(t)$ *(k > 0)*:
3. *Perform the semi-Lagrangian algorithm for all $\mathbf{x} \in \mathcal{T}^{(k)}(t)$:*
   a) *Calculate $\alpha(\mathbf{x})$ from (7.37).*
   b) *Calculate $\rho(\mathbf{x} - \alpha(\mathbf{x}), t - \Delta t)$.*
   c) *Update $\rho(\mathbf{x}, t)$ according to (7.36).*
4. *Estimate the local error $[\varepsilon]_\tau$ for each cell τ of $\mathcal{T}^{(k)}(t)$.*
5. *Refine those elements τ, where $[\varepsilon]_\tau > \theta_{\mathrm{ref}} \cdot [\varepsilon]_{\max}$, $[\varepsilon]_{\max} = \max_{\tau \in \mathcal{T}^{(k)}(t)} [\varepsilon]_\tau$, and θ_{ref} a given tolerance (see sect. 2.4). Coarsen the mesh analogously, to obtain $\mathcal{T}^{(k+1)}(t)$.*
6. *Set $k \leftarrow k + 1$.*
7. END WHILE

This adaptive semi-Lagrangian method has been introduced in [34]. We use cubic spline interpolation at the upstream position in the 2D case and a modified quadratic Shepard method in the 3D case [338]. In order to achieve

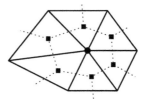

Fig. 7.3. Dual nodes (■), forming a dual cell, corresponding to an original node (•)

monotonicity of the solution density, we apply a clipping algorithm, as described in [34]. Algorithm 7.4.2 is defined pointwise, where we evaluate $\rho(\mathbf{x}, t)$ at grid points.

If we apply the same machinery to equation (7.33), we obtain

$$\int_{V(t)} \rho(\mathbf{x}, t) \, dx = \int_{V(t-\Delta t)} \rho(\mathbf{x}, t - \Delta t) \, dx, \tag{7.38}$$

where $V(t - \Delta t)$ is the upstream reference volume. So, instead of computing point values, we now need to define associated reference volumes $V(\mathbf{x}, t)$ for each grid point.

One possibility is to choose the dual cell corresponding to $\mathbf{x}$. The dual cell is the polyhedron, formed by the dual points $\xi_i(\mathbf{x})$, $i = 1 : n(\mathbf{x})$, surrounding $\mathbf{x}$ (see fig. 7.3). Here we take the surrounding cell's barycenters as dual points.

According to this definition of the reference volume, the upstream reference volume $V(\mathbf{x} - \alpha(\mathbf{x}), t - \Delta t)$ is then formed by the polyhedron of all upstream dual points $\xi_i^- := \xi_i(\mathbf{x}) - \alpha(\xi_i(\mathbf{x}))$. It is intuitively clear that we have to require the trajectory pieces $\alpha(\xi_i(\mathbf{x}))$ not crossing within one time-step, in order prevent volumes from collapsing and to guarantee stability of the method (Lipschitz-stability, see [371]). Now, we are able to define local and global mass conservation.

Definition 7.4.3 *(Global mass conservation)*
    Let $\Psi(\sigma)_{t_0}^{t_1} : \rho(\mathbf{x} - \alpha(\mathbf{x}), t_0) \rightarrow \rho(\mathbf{x}, t_1)$ *the discrete evolution equation implementing a numerical scheme denoted by* σ, $\mathbf{x} - \alpha(\mathbf{x})$ *being the upstream position, and suppose* $\mathcal{G} = \mathbb{R}^d$. *Let furthermore* $\mathbb{I} = [0, T]$. *Then we say that the scheme* σ *is globally mass conserving, if*

$$\int_{\mathcal{G}} \rho(\mathbf{x}, T) \, dx = \int_{\mathcal{G}} \Psi(\sigma)_0^T [\rho_0(\mathbf{x})] \, dx \overset{!}{=} \int_{\mathcal{G}} \rho_0(\mathbf{x}) \, dx.$$

Remark 7.4.4 *Note that the above definition can easily be extended to the case where* $\mathcal{G} \subsetneq \mathbb{R}^d$, $I = [t, T]$.

Definition 7.4.5 *(Local mass conservation)*

With the assumptions from 7.4.3 and denoting a volume corresponding to the space-time position $(\mathbf{x}, t)$ by $V(\mathbf{x}, t)$ we say that the scheme σ is locally mass conserving if

$$\int_{V(\mathbf{x},t)} \rho(\mathbf{x},t) \; dx = \int_{V(\mathbf{x},t)} \Psi(\sigma)_{t-\Delta t}^t (\rho(\mathbf{x}, t - \Delta t)) \; dx$$

$$\overset{!}{=} \int_{V(\mathbf{x}-\alpha(\mathbf{x}),t-\Delta t)} \rho(\mathbf{x}, t - \Delta t) \; dx. \tag{7.39}$$

Remark 7.4.6 *It is easy to see that if a scheme is locally mass conserving, then it is globally mass conserving, while the converse is not necessarily true.*

The locally mass conserving semi-Lagrangian schemes, described in the next section are all based on (7.39). So, they all discetize in one or the other way

$$\int_{V(\mathbf{x},t)} \rho(\mathbf{x},t) \; dx = \int_{V(\mathbf{x}-\alpha(\mathbf{x}),t-\Delta t)} \rho(\mathbf{x}, t - \Delta t) \; dx. \tag{7.40}$$

Note that (7.40) is the discrete counterpart of (7.38).

7.4.2 Mass-Conserving Algorithms

This section is devoted to the description of several locally mass conserving semi-Lagrangian schemes. The introduction to the algorithms is given for the 2D versions for simplicity. Where applicable, extensions to 3D are mentioned.

Quasi-Conservative Semi-Lagrangian Algorithm by Priestley

We briefly review this method, introduced by Priestley [327] and applied to shallow water equations by Gravel and Staniforth [177]. It is built on top of a quasi-monotone semi-Lagrangian scheme proposed by Bermejo and Staniforth [58]. This method is not locally mass conserving, but only globally mass conserving; and it is not based on equation (7.40).

The algorithm is based on a *blending* mechanism. Given a high order (possibly non monotone) and a low order (monotone) interpolation of the upstream concentration ρ^H and ρ^L resp., the overall upstream concentration ρ^- is calculated by the following formula:

$$\rho_k^- = \beta_k \rho_k^H + (1 - \beta_k)\rho_k^L$$

for each grid point $k = 1 : n$. β_k is maximized under the constraints $0 \le \beta_k \le 1$ and $\min([\rho_k, l], \rho_k^L) \le \rho_k^- \le \max([\rho_k, l], \rho_k^L)$, where $[\rho_k, l]$ is the set of the l neighboring concentration values. This describes essentially the algorithm of Bermejo and Staniforth.

Now, Priestley's algorithm adds another constraint to the β_k's, namely choose γ_k, $0 \le \gamma_k \le \beta_k$ with

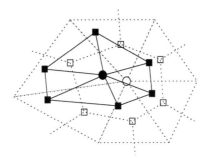

Fig. 7.4. The upstream dual cell triangulated (*solid lines*) overlayed over the original mesh (*with dotted dual cell as in fig. 7.3*)

$$\rho_k^- = \gamma_k \rho_k^H + (1 - \gamma_k)\rho_k^L$$

such that

$$\int_{\mathcal{G}} \rho^-(\mathbf{x})dx = \int_{\mathcal{G}} \rho^0(\mathbf{x})dx = C,$$

where ρ^0 is the initial condition function. Priestley's algorithm implements a linear programming approach and is very efficient. It has been implemented in the context of the adaptive semi-Lagrangian method in [34]. It adds a very small overhead to the computational work, while maintaining exact global conservation in the absence of convergent and divergent wind fields. However, it cannot maintain conservation properties in converging and accelerating wind.

Cell-Integrated Semi-Lagrangian Algorithm

The cell-integrated algorithm, introduced to the meteorology community by Machenhauer and Olk [271] and previously investigated by e.g. Semmazzi and Scroggs [360], is based on equation (7.40). In the mentioned approaches, based on a quadrilateral grid, each grid point is associated a cell. For each of the four cell corners the upstream position is determined. The concentration function is then integrated exactly over the upstream cell. Nair and Machenhauer and Nair, Scroggs and Semazzi give very efficient integration formulae based on geometric properties in [294, 295].

   Here, we extend the algorithm to locally refined triangular meshes. For simplicity, we consider $\rho(\cdot, t - \Delta t) = \bar{\rho}$ to be constant on grid cells. The aim is to compute $\rho(\mathbf{x}, t)$ for all grid points $\mathbf{x}$. Consider the upstream dual cell $V(\mathbf{x} - \alpha(\mathbf{x}), t - \Delta t)$ indicated by solid lines in fig. 7.4. The idea of the cell-integrated scheme is to geometrically intersect the upstream dual cell with the underlying fixed mesh. Then determine the mass contribution of each underlying cell to the (new) grid point value from the intersections.

   In mathematical terms, we triangulate the upstream dual cell $V(\mathbf{x} - \alpha(\mathbf{x}), t - \Delta t)$ into triangles $t_i^{(\mathbf{x}, t - \Delta t)}$, $i = 1 : l_\mathbf{x}$, $l_\mathbf{x}$ being the number of triangles forming the upstream dual cell (we use the index $\mathbf{x}$, since the coordinate determines uniquely the dual cell, the upstream dual cell, and their corresponding topological properties). Since we assume ρ to be constant on mesh cells τ the mass contribution to each $t_i^{(\mathbf{x}, t - \Delta t)}$ is given by

$$m_i = \sum_{\tau \in \mathcal{T}(t - \Delta t)} \left(|\tau \cap t_i^{(\mathbf{x}, t - \Delta t)}| \cdot \bar{\rho}|_\tau \right). \tag{7.41}$$

$\mathcal{T}$ is the mesh as in sect. 7.4.1. $\rho(\mathbf{x}, t)$ is then given by the formula

$$\rho(\mathbf{x}, t) = \frac{1}{|V(\mathbf{x}, t)|} \sum_i m_i. \tag{7.42}$$

We are left with two problems:

1. to find those elements $\tau \in \mathcal{T}(t - \Delta t)$, where $\tau \cap t_i^{(\mathbf{x}, t - \Delta t)} \neq 0$, and
2. to exactly determine the intersection $\tau \cap t_i^{(\mathbf{x}, t - \Delta t)}$.

The first problem can be solved efficiently by a recursive algorithm, provided that a mesh refinement hierarchy is available (i.e. a nested refinement strategy like in sect. 3.3 is used). Starting on the coarsest level, a recursive procedure enters into all those cells τ for which $\tau \cap t_i^{(\mathbf{x}, t - \Delta t)} \neq 0$. Once the recursion reaches the finest level, the contributions are collected and summed in a backward recursion.

Problem 2 can be solved by standard computational geometry algorithms. Since we only need to intersect triangles with triangles, this can be accomplished very efficiently as for example described in [9]. In sect. 4.5 we reported on the intersection capability of amatos.

Since the crucial part of this algorithm is the intersection, for a 3D version, we would need the intersection of convex polyhedra. However, this is a nontrivial task. The intersection in 2D is computationally much more demanding than e.g. the linear programming problem in the previous algorithm 7.4.2. In 3D we did not attempt to implement a cell-integrated scheme, since we anticipated computational overhead that were not acceptable.

Quadrature-Based Semi-Lagrangian Algorithm

Reviewing (7.40), we see that we have to solve two integrals numerically. While the cell-integrated method of the previous section achieves this goal by geometrically intersecting the upstream cell with the underlying grid, the quadrature based method represents the integrals by numerical quadrature formulae. Morton et al. point to a possible instability issue when combining semi-Lagrangian schemes with inexact quadrature in [292].

The idea is to find a quadrature rule for

$$\int_{V(\mathbf{x}, t)} \rho(\mathbf{x}, t) \, dx.$$

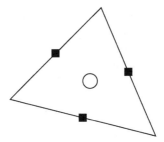

Fig. 7.5. Quadrature points for first ($\bigcirc$) and second ($\blacksquare$) order Newton-Cotes quadrature scheme

Since in an adaptive mesh the reference volume $V(\mathbf{x}, t)$ may be distorted, which is valid even more for the upstream volume $V(\mathbf{x} - \alpha(\mathbf{x}), t - \Delta t)$, we again triangulate the volume as shown in fig. 7.4.

Then, for each triangle in the dual cell, we can use a triangular quadrature scheme (see e.g. [138, 379]). For a first order approximation in each triangle, we can use

$$\int_{t_i} \rho(\mathbf{x}, t) \; d\mathbf{x} \approx |t_i| \cdot \rho(\xi_i, t),$$

where we denote t_i, $i = 1 : l_{\mathbf{x}}$ the triangles forming the reference volume $V(\mathbf{x}, t)$, ξ_i the center of gravity of t_i. A second order approximation is given by values on the triangle's edge centers, as indicated in fig. 7.5.

The quadrature based semi-Lagrangian scheme (with a first order quadrature scheme) is given by

$$\rho(\mathbf{x}, t) = \frac{1}{|V(\mathbf{x}, t)|} \sum_{i=1:l_{\mathbf{x}}} |t_i^{(\mathbf{x}, t - \Delta t)}| \cdot \rho(\xi_i - \alpha(\xi_i), t - \Delta t), \tag{7.43}$$

where $t_i^{(\mathbf{x}, t - \Delta t)}$ are the triangles forming the upstream dual cell. Higher order quadrature extends the above formula in an obvious way.

The quadrature-based algorithm by construction is not exactly (i.e. up to machine precision) mass conserving, because the values for ρ at quadrature points are still interpolated and include error. However, this method is locally mass conserving up to the order of discretization, provided a quadrature rule of at least discretization order is used.

This scheme can be naturally extended to 3D. While the upstream dual cell now becomes a polyhedron in 3D, there is a triangulation into tetrahedral elements for which Newton-Cotes cubature formulae are given in [138]. The computational effort, however, is increased over the 2D case, since a lot more tetrahedra are needed to triangulate the upstream dual cell, resulting in an increased number of cubature points, and the interpolation at cubature points in 3D is computationally more expensive.

Cell-Weighted Semi-Lagrangian Algorithm

The idea for the cell-weighted semi-Lagrangian algorithm is based on a simplification of the first order quadrature scheme just introduced. Here, we simply approximate the integral by

$$\int_{V(\mathbf{x},t)} \rho(\mathbf{x},t) \; dx \approx |V(\mathbf{x},t)| \cdot \rho(\mathbf{x},t).$$

Doing so on both sides of (7.40) results in the following simple scheme

$$\rho(\mathbf{x},t) = \frac{|V(\mathbf{x} - \alpha(\mathbf{x}), t - \Delta t)|}{|V(\mathbf{x},t)|} \cdot \rho(\mathbf{x} - \alpha(\mathbf{x}), t - \Delta t). \qquad (7.44)$$

Thus, the plain semi-Lagrangian scheme is modified by a factor $F = \frac{|V(\mathbf{x}-\alpha(\mathbf{x}),t-\Delta t)|}{|V(\mathbf{x},t)|}$, with $|\cdot|$ the area or volume, that recovers the conservation property.

This algorithm is very efficient, since additionally to the plain semi-Lagrangian method, it only needs to compute two areas/volumes. It can be extended easily to 3D. Note that we did not use any 2D specific notion in the formulation of the algorithm. However, as in the previous section, this algorithm is not exactly mass conserving (up to machine precision), since there is still an interpolation involved. For a constant, homogenous wind field, this algorithm reduces to the plain semi-Lagrangian algorithm, because $F = 1$ then. A similar approach can be found in [320].

Mass-Packet Based Semi-Lagrangian Algorithm

To describe the Mass-Packet Semi-Lagrangian Method (MPSLM) [285], we need to define mass packets. These are mass-volume units which are defined by their mass, volume and barycenter's position. In general they are much smaller than the grid cells. They are used to subdivide the cells into smaller parts. The mass packets, created in each time step, will be advected according to the semi-Lagrangian idea. In this section, we will denote an item in the old mesh $\mathcal{T}(t - \Delta t)$ by superscript $^{(-)}$ while a new item at time t is denoted by superscript $^{(+)}$.

First, the density $\rho^{(-)}$ must be transformed to mass values. Therefore, we define a mass attached at each node $N_j^{i(-)}$ per cell $\tau_j^{(-)}$ by

$$m(N_j^{i(-)}) = \rho(N_j^{i(-)})|A_j^i|, \qquad (7.45)$$

where we denote the volume by $|\cdot|$, $|A_j^{i(-)}|$ is the corresponding area of $\tau_j^{(-)}$ as depicted in fig. 7.6.

Then the mesh cell $\tau_j^{(-)}$ is fully partitioned into L disjoint pieces: $\sum_{l=1}^{L} |M_j^l| = |\tau_j^{(-)}|$ where the M_j^l are the L mass packets of cell $\tau_j^{(-)}$ (see

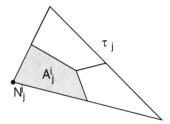

Fig. 7.6. Volume $|A_j^i|$ in Cell τ_j is associated to node N_j^i

fig. 7.7). According to the mass packet's volume $|M_j^l|$ and barycenter's position λ_i^l with respect to the cell's nodes $N_j^{i(-)}$, packets get mass assigned by the formula

$$m(M_j^l) = \sum_{i=1}^{d+1} \lambda_i^l m(N_j^{i(-)})|M_j^l| \tag{7.46}$$

Note that the barycentric coordinates have the property that $\sum_l \lambda^l = 1$. Thus, keeping in mind the disjoint partition of cells in mass packets, summing over all mass packets in a cell, we obtain the exact mass assigned to each node. By this means the mass in mesh $\mathcal{T}(t - \Delta t)$ is virtually transformed into mass packets.

For each node of the new grid the upstream position is calculated following the plain SLM (7.36). Knowing the upstream grid cells and the mass packets on the old grid, a mapping step follows. The mapping consists of assigning each upstream (new) cell $\tau_j^{(+)}$ the corresponding (old) mass packets. A mass packet is assigned to cell $\tau_j^{(+)}$ if its barycenter is inside the upstream position of $\tau_j^{(+)}$. It should be noted that this search algorithm represents the most time consuming part of the whole scheme. The mapping is illustrated in fig. 7.8.

After mapping the mass packets to new cells, the barycentric coordinates μ_i^l with respect to the (new) nodes $N_j^{i(+)}$ are computed. Finally, the nodes

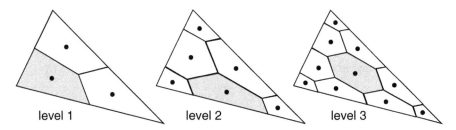

level 1   level 2   level 3

Fig. 7.7. Mass packet refinement within one cell. Different mass packet types are shaded. The number of mass packets is given by $L = \sum_{i=1}^{\text{level}+1} i$. The adaptive mesh refinement requires an adaptive number of mass packets, since we require a continuous mapping of mass to the upstream cells

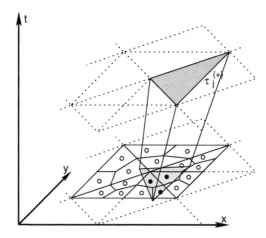

Fig. 7.8. Mass packet mapping: The (*solid*) mass packets are associated to the upstream position of $\tau_j^{(+)}$

$N_j^{i(+)}$ get mass assigned corresponding to the mass packets found in cell $\tau_j^{(+)}$:

$$m(N_j^{i(+)}) = \sum_{l:M_j^l \in \tau_j^{(+)}} \mu_i^l m(M_j^l) \tag{7.47}$$

As for the λ's it holds that $\sum_l \mu^l = 1$. Therefore, the mass is reassigned to the nodes without loss, except for mass packets lying outside the upstream (new) mesh.

Now that we know the mass at the (new) nodes, we can compute the density value

$$\rho(N_j^{i(+)}) = \frac{m(N_j^{i(+)})}{|V^{i(+)}|} \tag{7.48}$$

where $V^{i(+)}$ is the control volume associated to node $N^{i(+)}$ composed of all A_j^i in the patch of the node. This scheme guarantees that ρ is non-negative. Details of the implementation and theoretical proofs of the scheme's conservation properties can be found in [285]. Summing up, the MPSLM algorithm is given by

Algorithm 7.4.7 (*Mass Packet semi-Lagrangian Algorithm*)

1. *Compute nodal masses $m(N_j^{i(-)})$ using (7.45).*
2. *Subdivide cells into mass packets and assign masses according to (7.46).*
3. *Compute upstream positions to $\tau_j^{(+)}$ and map mass packets to $\tau_j^{(+)}$.*
4. *Compute new nodal masses using (7.47).*
5. *Reconstruct nodal densities with (7.48).*

8

Example Applications

This chapter is devoted to some example applications in adaptive atmospheric (and oceanic) modeling. Of course, there are many more examples of the application of adaptivity in atmospheric modeling. These examples represent projects that were realized by or in collaboration with the author. The first two examples in sections 8.1 and 8.2 are linear tracer transport examples, the third and fourth examples (section 8.3) are extracted from the research project PLASMA in which dynamical cores for a simplified climate model are under development. While these four examples share semi-Lagrangian time-discretization schemes which are well suited for adaptive mesh refinement (see [34]), the fifth example, demonstrating adaptive wave dispersion in section 8.4, employs a finite volume type discretization method for the shallow water equations.

8.1 Tracer Advection

Tracer advection – though very simple in it's set-up – is a good testing environment for adaptive numerical schemes and for adaptive concepts. This is the reason, why some detailed investigation of this simple application is presented in this section. The example is a 2D tracer dispersion application in the arctic stratosphere. The following material is taken from a publication by the author and coworkers [39].

8.1.1 Filamentation of Trace Gas in the Arctic Polar Vortex

The perturbed chemistry within the polar stratospheric vortex has generated great interest in the respective dynamics. The atmospheric distribution of trace constituents in the lower stratosphere not only depends on chemical sources and sinks but also on the redistribution as a result of transport, induced by various dynamical processes [409].

The quasi-horizontal dispersion of a passive tracer on synoptic time scales is likely to occur through the formation of tracer filaments in the presence of a background wind shear. Plumb et al. showed that fine structures seen in aircraft data coincide with filamentary structures produced by a contour advection model [323]. Edouard et al. claim that the ozone depletion depends on the resolution of the model, as small scale structures influence the local depletion rate [136]. Although these results have been questioned in [361], it seems to be common sense that model resolution has an impact on density distribution and therefore on reaction rates. The scale cascade of tracers under the influence of shear and strain has also been investigated in detail by Haynes and Anglade [191].

Bregman et al. carried out in situ measurements of the trace gases O_3, HNO_3, and N_2O in the Arctic lower stratosphere during February 1993 on board of a Cessna Citation aircraft during the first Stratosphere-Troposphere Experiment by Aircraft measurements (STREAM) campaign [72]. Strong variations in the concentrations and distributions of these trace gases were found. The time series with a resolution of minutes showed pronounced small-scale variations and large horizontal variations in the range of some dozen kilometers, indicating that the aircraft flew through air masses with different origins. Unfortunately these data do not allow the identification of horizontal scales below some dozen kilometers.

The horizontal resolution required for the reproduction of filaments cannot be achieved by general circulation models to date. Most approaches, therefore, use offline chemistry transport models (CTM) with high local resolution. When attempting to model filamentary structures, most other investigators use contour advection models (see [130] for a description). Other approaches utilize purely Lagrangian particle advection schemes (see e.g. [378]). In this section, we adopt an adaptive modeling technique, developed in [34], and [36], in order to achieve very high local resolution. (we go as far as approx. 5 km here).

Three main aspects are investigated.

- The influence of varying resolution of wind data;
- The applicability of very-high-resolution advection to coarse-resolution wind data;
- The influence of varying horizontal resolution of the transport model on the formation of small scale filamentary structures.

Other studies investigated the influence of resolution (see [408]).

We consider the horizontal transport of a conserved passive tracer in the lower polar stratosphere in an idealized configuration, using the wind components from a high-resolution regional climate model (HIRHAM) [119, 340]. HIRHAM is a sophisticated climate model, which resolves the arctic region with a horizontal resolution of $0.5°$. It is forced at the lateral and lower boundaries by ECMWF reanalysis data. The simulated geopotential structure at 73.4 hPa (we call this the 70 hPa layer) during January 1990 with a cold low

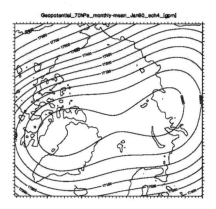

Fig. 8.1. January monthly mean geopotential (m) for the Arctic, simulated in HIRHAM for 1990

is presented in fig. 8.1, well known as Polar Vortex with, in our situation two centers, one north of Greenland the other south-east of Novaya Zemlya.

8.1.2 Adaptive Tracer Transport Model

For the experiments, described in section 8.1.3, we use an adaptive tracer transport model. While a detailed description of the basic model ideas can be found in [34], we give a brief overview here. The model solves the advection equation in two dimensions:

$$\partial_t \rho + \nabla \cdot (\mathbf{v}\rho) = R \tag{8.1}$$

The wind $\mathbf{v} = \mathbf{v}(\mathbf{x}, t)$ is given by the HIRHAM data (cf. section 8.1.1). $R = R(\mathbf{x}, t)$ is the right hand side of (8.1) which can contain additional forces, but is set to zero in our case (i.e. we simulate the advection of an idealized passive tracer without diffusion and without other sources and sinks). $\rho = \rho(\mathbf{x}, t)$ denotes the (scalar) concentration density of the tracer, while $\mathbf{x} = (x, y)$ is the horizontal coordinate in the 70 hPa layer.

The time dependent part of the equations is discretized by the conservative cell-weighted semi-Lagrangian method (see section 7.4.2). An adaptive grid generator (`amatos`) refines the triangular grid down to a lower bound, where the gradient of the tracer concentration $\nabla \rho$ is large. On the other hand, in regions where $\nabla \rho$ is small, the grid is coarsened up to an upper bound. The maximum-norm based refinement strategy is used (see algorithm 2.4.1 in section 2.4). We set $\theta_{\text{ref}} = 0.2$ and $\theta_{\text{crs}} = 0.1$. Table 8.1 gives the grid type according to the classification in section 3.2 and the corresponding horizonal grid resolution used in the experiments.

Table 8.1. Highest local grid resolution and corresponding grid type for the adaptive simulation

grid type	horiz. resolution (km)
CNT[12]	55
CNT[4,15]	20
CNT[4,17]	10
CNT[4,19]	5

With this mechanism, we achieve very high local resolution without exhausting limited computing resources. An example of an adaptively refined grid is given in fig. 8.2.

8.1.3 Experiments and Results

In this sub-section we try to answer three main questions. The first one is of rather philosophical character, as we ask: why can we see small scale filaments in the tracer advection when these structures are not present (due to coarse resolution) in the driving wind data?[1] There are other more in depth investigations of this question [323, 191], however we decided to take this example into consideration because of its simple and instructive character. The more practical aspects of our investigation will analyze the influence of resolution of wind data and model resolution to the formation of small scale structures.

Influence of wind gradient

In order to explore the first question, let us think of a simplified example: let $\mathbf{v}$ in (8.1) be a constant circular wind with decreasing velocity towards the boundary, acting in a closed box: $\mathbf{v}(x,y) = [u(x,y), v(x,y)] := [-y \cdot d, x \cdot$

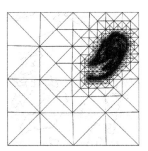

Fig. 8.2. Adaptively refined grid, corresponding to the situation in fig. 8.6

---

[1] In fact, sub-grid features are present in the wind data as they are parameterized in the underlying model and thus contribute to the shape of the wind field!

Fig. 8.3. Initial configuration for the model problem with a circular wind field

d], where $d := 4 \cdot \|(x_{center}, y_{center}) - |(x, y)|\|_2$. Let the initial concentration distribution be given by a rectangle with concentration 1 and zero everywhere else (see fig. 8.3).

After several time steps one will observe formation of small scale stirring in the transport model, while obviously no dynamical features are present in the underlying data. These small scale structures are caused by the presence of a gradient in the wind field acting orthogonally to the gradient of the concentration. In general, whenever a gradient in the wind field is non-tangential to the gradient of the concentration, stirring will occur. If this process can be resolved fine enough by the transport scheme, we will see small scale filamentation. On the other hand, if resolution is too coarse, filaments will not be visible due to numerical dissipation (see fig. 8.4).

It is therefore reasonable to use a very-high-resolution tracer transport scheme with data that are not equally highly resolved. The wind data contain sub-grid scale information, due to parameterized phenomena. The shape of the wind field is influenced by these parameterizations, resulting in filamentation by the above mentioned interaction of gradients.

Influence of Wind Data Resolution

In order to determine the influence of data resolution (i.e. the resolution of the given wind field) to the formation of small scale filaments, wind data are given on grids of 110×100, 55×50, and 22×20 grid-points respectively. This corresponds to a mesh size of approx. 50 km, 100 km, and 250 km respectively (or approx. $0.5° \times 0.5°$, $1° \times 1°$, and $2.5° \times 2.5°$ resp.). Minimal mesh size of the transport model is set to approx. 10 km, corresponding to the CNT[6,17] mesh. Here and in all the following experiments, the tracer concentration is initialized as shown in fig. 8.5. This position corresponds to the left "eye" of the Arctic Polar Vortex in January 1990.

Compared to the reference case (110×100 data points), the 55×50 grid-point case shows no significant changes. However, reducing the data field to only 22×20 points results in a different behavior. On the one hand, wind structure is slightly different, resulting in a displaced path of the transport.

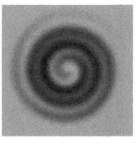

Fig. 8.4. Result of the advection with the model problem after some hours of model-time. With high (local) resolution, fine structures are visible (**left**) while numerical dissipation has led to heavy erosion with coarse (global) grid resolution (**right**). Note that the total mass of the tracer is conserved in both cases

Fig. 8.5. Initial tracer distribution for the experiments

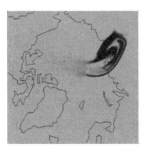

Fig. 8.6. Tracer concentration after 288 hours of model time, using highly resolved wind data on a grid of 110×100 grid-points (**left**), versus low-resolution 22×20 grid-points (**right**)

On the other hand, a more interesting behavior, namely the formation of small scale filaments, is deteriorated. When comparing both plots in fig. 8.6, one observes that fine stirring structures are present only in the high-resolution case.

Note that interpolation of the wind data set to the computational grid of the very-high-resolution transport model is a problem in itself. We tried bi-linear and bi-quadratic interpolation and found no significant difference. In the experiments we use bi-linear interpolation for efficiency and shape preserving reasons. This is in accordance with the choice in the contour advection simulations in [408].

Influence of Model Resolution

The next series of experiments aims at determining a necessary computational grid resolution for the formation of small scale filaments. Wind data are taken from the high-resolution data set (i.e. 110×100 data-points). The first case uses a uniformly refined grid with a resolution of approx. 55 km. This is almost the resolution of the given data set. Three additional model configurations are probed with local resolutions of approx. 20 km, 10 km, and 5 km respectively (see table 8.1). Note that all of these high-resolution tracer advection simulations can be easily computed on a standard workstation due to the adaptive refinement strategy.

As one would expect, finer resolution causes finer structures in the tracer concentration function. Figure 8.7 shows the tracer concentration for the different resolutions. While there are more and more second order filamentary structures visible, the main path and dispersion of the tracer field remain unchanged when increasing the resolution.

8.2 Inverse Tracer Advection

In air quality modeling and monitoring, one is often faced with the problem of reconstructing a source of pollution from measurement data. So, the inverse problem (of a forward directed simulation of tracer dispersion) can be formulated as follows:

Given a tracer (density) distribution $\rho(\mathbf{x}, T)$ at some time T, what was the distribution at time $t < T$?

Note that the time advancement faces into the negative direction. This problem can be solved by the adjoint operator to the advection operator [137]. The adjoint or inverse modeling mode is sometimes called backward mode [362]. There is a large resource of meteorological and especially oceanographical literature on adjoint modeling or assimilation [260, 164, 357, 385]. Many of the adjoint models are derived by automatic differentiation techniques, so called adjoint compilers or pre-processors [99, 180]. Sirkes and Tziperman however,

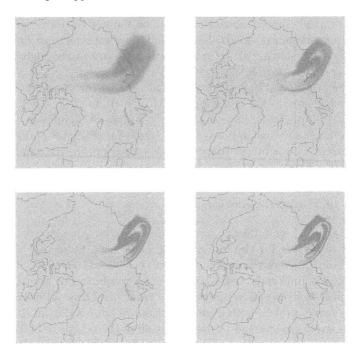

Fig. 8.7. Tracer concentration after 288 hours of model time, computed on a uniform grid with approx. 55 km resolution (**top, left**) CNT[12], and computed on an adaptive grid with approx. 20 km (**top, right**) CNT[4 : 15], 10 km (**bottom, left**) CNT[4 : 17], and 5 km (**bottom, right**) CNT[4 : 19] resolution respectively

argue that this (automated) approach might not be numerically stable, since it is not necessarily true that a stable forward discretization is also stable after automatic differentiation in the backward direction [366]. We do not comment on this result, but take it as a good reason for deriving the adjoint equations analytically and discretizing these equations by a numerically stable adaptive method. We shall briefly describe this technique here, referring to [331]. The reader is also referred to [386]. A similar strategy, utilizing a cost function and the generic finite element weak formulation of the model equations has been used by Dobrind for inversion of an ocean model [126].

Let $\mathcal{D}$ be the linear differential operator, defined by

$$\mathcal{D}(\cdot) = \partial_t(\cdot) + \nabla \cdot (\mathbf{v}(\cdot)), \tag{8.2}$$

with a given velocity field $\mathbf{v}$. Then the advection equation, given by

$$\mathcal{D}(\rho) = r, \tag{8.3}$$

represents the transport of a tracer with a source term $r = r(\mathbf{x}, t)$. For educational reasons and because in atmospheric simulation (lateral) boundary conditions do not play a dominant role, we omit the detailed specification of adequate boundary and initial conditions and assume that these are suitably given. Furthermore, we assume that ρ is given on a domain $\overline{\mathcal{G}} = \mathcal{G} \times I \subset \mathbb{R}^d \times \mathbb{R}$, $I = [0, T]$. Let $H = H^1(\overline{\mathcal{G}})$ be the Sobolev space of square integrable functions in $\overline{\mathcal{G}}$. Then let $\mathcal{D}$ be defined on the subset $D(\mathcal{D}) \subset H$ of functions that satisfy the boundary and initial conditions[2].

The adjoint operator $\mathcal{D}^*$ is defined by the duality relation

$$\langle \mathcal{D}\phi, \psi \rangle = \langle \phi, \mathcal{D}^*\psi \rangle, \tag{8.4}$$

where $\phi \in D(\mathcal{D})$ is an arbitrary function in the primal function space, $\psi \in D(\mathcal{D}^*) \subset H$ is an arbitrary function in the dual space, and $\langle \cdot, \cdot \rangle$ is the inner product on H, given by

$$\langle \phi, \psi \rangle = \int_0^T \int_{\mathcal{G}} \phi\psi \, dxdt,$$

with $\phi, \psi \in H$. Note that $D(\mathcal{D}^*)$ is different from $D(\mathcal{D})$, and chosen such that the adjoint operator is well defined with appropriate boundary conditions. Now, for $\phi = \rho$ and $\psi = \rho^*$ we derive

$$\langle \mathcal{D}\rho, \rho^* \rangle = \langle \rho, \mathcal{D}^*\rho^* \rangle. \tag{8.5}$$

From (8.5) we deduce in analogy to (8.3) the adjoint advection equation given by

$$\mathcal{D}^*\rho^* = q, \tag{8.6}$$

with q representing a function of measurements, according to the problem formulation in the beginning of this section. By substitution of (8.6) and (8.3) into (8.5) it is easy to see that

$$\langle r, \rho^* \rangle = \langle \rho, q \rangle.$$

In order to derive an explicit formula for the adjoint equation to the advection equation, we proceed straight forward by substituting (8.2) into (8.5) to obtain

$$\langle \mathcal{D}(\rho), \rho^* \rangle = \int_0^T \int_{\mathcal{G}} \partial_t\rho\rho^* \, dxdt + \int_0^T \int_{\mathcal{G}} \nabla \cdot (\mathbf{v}\rho)\rho^* \, dxdt.$$

After integration by parts and applying Green's theorem, we get

$$\langle \mathcal{D}(\rho), \rho^* \rangle = -\int_0^T \int_{\mathcal{G}} \rho\partial_t\rho^* \, dxdt - \int_0^T \int_{\mathcal{G}} \rho\mathbf{v} \cdot \nabla\rho^* \, dxdt + F,$$

---

[2] We deliberately omit the details of Hilbert spaces, specification of boundary and initial conditions, etc. and accept a great portion of sloppiness in this section. It is our intention to introduce to the basic idea of the adjoint and not to formulate the mathematical foundations of the derivation.

where F represents further terms that arise from the integration by parts. Careful observation reveals that $F = 0$ for suitably chosen boundary conditions on ρ^* and ρ. Therefore, we derive, using again the duality relation (8.4)

$$\langle \rho, \mathcal{D}^*(\rho^*) \rangle = \int_0^T \int_{\mathcal{G}} \rho\left(-\partial_t \rho^* - \mathbf{v} \cdot \nabla \rho^*\right) \, dx dt. \tag{8.7}$$

From this, we deduce the adjoint operator to the advection operator given in (8.2):

$$\mathcal{D}^*(\cdot) = -\partial_t(\cdot) - \mathbf{v} \cdot \nabla(\cdot). \tag{8.8}$$

Assuming a divergence free wind field ($\nabla \cdot \mathbf{v} = 0$), homogeneous forcing ($r \equiv 0$ and thus $q \equiv 0$), and a semi-Lagrangian advection scheme, where we discretize the material derivative

$$\frac{d\rho}{dt} \approx \frac{\rho(x,t) - \rho(x-\alpha, t-\Delta t)}{\Delta t}$$

where α denotes the upstream particle path, we see that solving the adjoint equation is equivalent to solving the forward equation with an inverted (negative) wind field.

An academic example for the derived adjoint model is given in fig. 8.8. If we take the wind field for a forward computation as in the circular wind test case in section 8.1.2 above and the inverse wind being the negative of that. Then, we can

1. initialize the scalar field $\rho|_{t=0}$ by an arbitrary function (e.g the logo "TUM");
2. compute the (discrete) solution at time T, $\rho_h|_{t=T}$ by a forward integration;
3. take $\rho^{\mathrm{inv}}|_{t=T} = \rho_h|_{t=T}$ as initial data for the inverse (adjoint) model experiment,
4. integrate backwards in time to solve for $\rho_h^{\mathrm{inv}}|_{t=0}$, which should correspond to $\rho|_{t=0}$.

We only show steps 3 and 4 in fig. 8.8. It can clearly be seen that adaptivity helps to reduce the diffusive error considerably. This is just a case study and further investigation needs to be conducted.

8.3 Shallow Water Equations

This section relies mostly on the results of the project *PLASMA*, which is a joint research project of the author's group with two groups at the branches of Alfred-Wegener-Institute in Bremerhaven and Potsdam. Two paths have been opened for the development of a new adaptive dynamical core of an atmospheric model. A vorticity-divergence formulation and a velocity vector formulation. The first form has been tested extensively, while the second form has just been released (which is the reason for the relatively brief description).

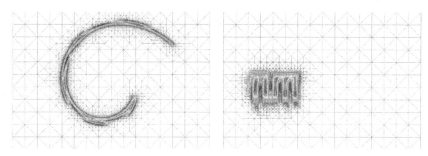

Fig. 8.8. Initial field of tracer distribution (with grid) and final inverse modeling result after 48 hours of model time

8.3.1 Vorticity-Divergence Formulation

The implementation in vorticity-divergence form is called *PLASMA-P*. While the formulation of the model equations and their discretization has been developed mainly by Matthias Läuter [255, 258, 332], the adaptive mesh technology is provided by the package `amatos` [44], developed by the author's group and the solution of the large linear systems of equations is comprised with the help of the solver library FoSSI by Frickenhaus and coworkers [156].

PLASMA-P in the current 2D version solves the spherical (rotational) shallow water equations (see appendix C) in vorticity-divergence formulation. The starting point for the derivation is given by the vectorial shallow water equations:

$$\frac{d\mathbf{v}}{dt} + \nabla_s \Phi + f_C \mathbf{n} \times \mathbf{v} + |\mathbf{v}| <^2 \mathbf{n} = 0,$$
$$\frac{d\Phi}{dt} + (\Phi - \Phi_0)\nabla_s \cdot \mathbf{v} - \mathbf{v} \cdot \nabla_s \Phi_0 = 0, \qquad (8.9)$$
$$\mathbf{v} \cdot \mathbf{n} = 0.$$

Note that we used the tangential material derivative

$$\frac{d}{dt} = \partial_t + \mathbf{v} \cdot \nabla_s.$$

These equations have been derived, starting from the shallow water equations as given in [100]. Remember that we denoted by ∇_s the tangential (horizontal) ∇-operator on the sphere (see appendix C). Φ denotes the geopotential height and Φ_0 the orography, while $\mathbf{v}$ is the velocity and $\mathbf{n}$ the unit outward normal on the sphere.

We define the vorticity $\zeta = \nabla_s \times \mathbf{v}$ and the divergence $\delta = \nabla_s \cdot \mathbf{v}$. Applying the rotation and divergence operators to the momentum equation (first equation in (8.9)), a scalar formulation of the shallow water equations can be obtained, where we get additional equations from the Helmholtz decomposition.

$$\frac{d\zeta}{dt} + \zeta\delta + \delta f_C + \mathbf{v} \cdot \nabla_s f_C = 0$$

$$\frac{d\delta}{dt} + \Delta_s \Phi - \zeta f_C + (\mathbf{n} \times \mathbf{v}) \cdot \nabla_s f_C + J(\mathbf{v}) = 0$$

$$\frac{d\Phi}{dt} + \delta(\Phi - \Phi_0) - \mathbf{v} \cdot \nabla_s \Phi_0 = 0 \qquad (8.10)$$

$$-\Delta_s \psi = \zeta$$

$$\Delta_s \chi = \delta$$

$$\nabla_s \times \psi + \nabla_s \chi = \mathbf{v}.$$

The functions $\psi, \chi : S \times [0, T] \to \mathbb{R}$ are the *stream function* and *velocity potential* respectively, and J is defined by

$$J(\mathbf{v}) = (\nabla_s v_i)^T A^{i,j} (\nabla_s v_j) + \mathbf{v} \cdot \mathbf{v},$$

with $A^{i,j}_{k,l} = \delta_{il}\delta_{jk}$, and $\delta_{\mu\nu}$ the Dirac delta function. The first three equations in (8.10) are the actual scalar shallow water equations, while the last three equations represent the Helmholtz decomposition of the vector field $\mathbf{v}$. A thorough derivation and proof for the equivalence of (8.9) and (8.10) can be found in [256].

Equations (8.10) are discretized by an implicit Lagrange-Galerkin approach. However, the process is a little more complicated and is only outlined here. Again the details are given in [256]. In fact, PLASMA-P does not solve (8.10) but uses a stabilization by diminishing numerical diffusion. Thus, the advection operators $\mathbf{v} \cdot \nabla_s \varphi$, where $\varphi = \zeta, \delta, \Phi$ (hidden in the material derivative operators) are replaced by the operator $\mathbf{v} \cdot \nabla_s \varphi - \nu \Delta_s \varphi$ with suitably chosen diffusion parameter ν.

A semi-discretization of (8.10) is given by the following system. Note that the mixed terms $\zeta\delta$ and $\Phi\delta$ have been treated by a time-averaging. Furthermore we have written $d_t\varphi = \frac{d\varphi}{dt}$.

$$(d_t\zeta, b_h) + \nu_1(\nabla_s\zeta, \nabla_s b_h) + (\zeta\delta, b_h) + (f_C\delta, b_h) = (F_1, b_h)$$

$$(d_t\delta, b_h) + \nu_1(\nabla_s\delta, \nabla_s b_h) + (\nabla_s\Phi, \nabla_s b_h) - (f_C\zeta, b_h) = (F_2, b_h)$$

$$(d_t\Phi, \beta_h) + \nu_2(\nabla_s\Phi, \nabla_s\beta_h) + (\delta\Phi, \beta_h) - (\Phi_0\delta, \beta_h) = (F_3, \beta_h) \quad (8.11)$$

$$(\nabla_s\psi, \nabla_s\gamma_h) = (\zeta, \gamma_h)$$

$$(\nabla_s\chi, \nabla_s\gamma_h) = -(\delta, \gamma_h)$$

$$(\nabla_s \times \psi + \nabla_s\chi, \mathbf{w}_h) = (\mathbf{v}, \mathbf{w}_h).$$

Here, b_h, β_h, γ_h, and $\mathbf{w}_h$ represent suitable variational test functions, while $(\cdot, \cdot)$ is the appropriate inner product; we also assumed suitably chosen Hilbert spaces for the equations to be well defined. Now the first three equations in (8.11) are further discretized in time by a semi-Lagrangian time integration method. Using quadrature rules to discretize the inner products above, finally leads to a large blocked system of equations, given by

$$\begin{bmatrix} M_1 + \Delta t(\nu_1 S_1 + B_1) & \Delta t(C + B_2) & 0 \\ -\Delta t C & M_1 + \Delta t \nu_1 S_1 & -\Delta t S_{12} \\ 0 & B_4 & M_2 + \Delta t(\nu_2 S_2 + N_3) \end{bmatrix} \begin{bmatrix} \zeta_h \\ \delta_h \\ \Phi_h \end{bmatrix} = \begin{bmatrix} F_1 \\ F_2 \\ F_3 \end{bmatrix}.$$

(8.12)

The vectors ζ_h, δ_h, and Φ_h represent the coefficients of vorticity, divergence and geopotential height field in the Galerkin sense. For a variable φ, $\varphi_h = (\varphi_{h,1}, \ldots, \varphi_{h,n})$ denotes the vector of coefficients of the Galerkin representation

$$\varphi(\mathbf{x}) = \sum_{i=1:n} \varphi_{h,i} b_{h,i}(\mathbf{x})$$

of the function φ corresponding to basis functions b_h. Furthermore, we denote by M_i the mass matrix, S_i the stiffness matrix, C the Coriolis matrix with $c_{i,j} = (f_C b_i, b_j)$, and B_i appropriately defined mixed variants of the mass matrix. Finally F_i are the right hand sides obtained by multiplication with the mass matrix. The Helmholtz decomposition related equations in (8.11 are discretized in a similar manner by the finite element method. To summarize, the solution process in PLASMA-P is given by the following algorithm.

Algorithm 8.3.1 *(Discretization of PLASMA-P)*
Let initial values for $\mathbf{v}$, *and* Ψ *be given.*

1. *Initialize* ζ *and* δ *from the data.*
2. *Compute trajectories by solving the ordinary differential equation*

$$\dot{\mathbf{x}} = \mathbf{v}(\mathbf{x}, t)$$

   for each grid point (see section 7.4.1).
3. *Compute the right hand sides in (8.12), using the trajectories.*
4. *Solve the system of equations (8.12).*
5. *Solve the Poisson problems*

$$S_2 \psi_h = M_{21} \zeta_h, \quad S_2 \chi_h = M_{21} \delta_h,$$

   and derive the new velocity vector $\mathbf{v}_h$ *from*

$$M_3 \mathbf{v}_h = R \psi_h + G \chi_h,$$

   where R is the rotation matrix and G the divergence matrix from the finite element discretization, resp.

In this scheme, the prognostic variables are the geopotential Φ, the vorticity ζ, and the divergence δ. This is convenient, since these are all scalar values and secondly, the divergence and vorticity can immediately be used as refinement criteria in the following L^2-sense:

$$\eta_\tau = \int_\tau \delta^2 + \zeta^2 \, dx.$$

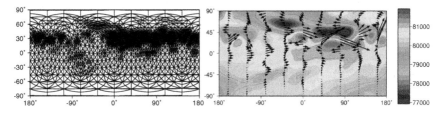

Fig. 8.9. Simulation of Rossby waves with PLASMA-P. A zonal jet, located at $30°N$ with wind maximum of $40ms^{-1}$ is disturbed by a $7km$ high mountain, located at $30°N$, $0°E$. (**Left**), grid (**right**) geopotential height field (in m^2s^{-2}, right) after 20 days (grid type $CNT(1041:428)_{[km]}$)

The rationale behind this choice is that a flow is less well approximated in areas of high turbulence. Now, turbulence occurs in areas of high vorticity or divergence (in first order approximation).

PLASMA-P has been validated mainly with the help of analytical test cases, described in section 8.5. Figure 8.9 shows the adaptively refined grid and a snapshot for a test case with a jet over an isolated mountain.

8.3.2 Velocity Formulation

The development branch building on the velocity formulation of the shallow water equations is called *PLASMA-FEMmE*, since it builds on the development described in [193, 194]. The basic formulation of the shallow water equations employed by PLASMA-FEMmE is given by

$$\frac{d\mathbf{v}_s}{dt} + \nabla_s \Phi + f_C(\mathbf{e}_r \times \mathbf{v}_s) = 0, \qquad (8.13)$$

$$\frac{d\Phi}{dt} + \Phi\nabla_s \cdot \mathbf{v}_s = 0. \qquad (8.14)$$

These equations are discretized in time by a semi-implicit semi-Lagrangian 2-time-level scheme introduced in [388]. This 2-time-level scheme has been derived from a 3-time-level scheme, therefore some of the terms have to be evaluated at the midpoint of the trajectory. We denote by:

- φ^+ a constituent φ's value at a grid point at an unknown time $t + \Delta t$,
- φ^- a constituent φ's value at an upstream position at a known time t, and
- φ^* a constituent φ's value at a trajectory midpoint extrapolated to time $t + \Delta t/2$.

With this notation, and a splitting of the difference of free surface of geopotential height Φ and orography Φ_H (i.e. the effective geopotential height) into a constant basic state $\overline{\Phi}$ and a disturbance $\widehat{\Phi}$,

$$\widehat{\Phi}(\mathbf{x}, t) + \overline{\Phi} = \Phi(\mathbf{x}, t) + \Phi_H(\mathbf{x}),$$

we write the semi-Lagrangian time discretization of (8.13) and (8.14) by

$$\frac{\mathbf{v}^+ - \mathbf{v}^-}{\Delta t} = - \left[\omega \nabla_s (\Phi^+ + \Phi_H^+) + (1 - \omega) \nabla_s (\Phi^- + \Phi_H^-) \right] -$$
$$- \left[f_C (\mathbf{e}_r \times \mathbf{v}) \right]^*, \tag{8.15}$$

$$\frac{\widehat{\Phi}^+ - \widehat{\Phi}^-}{\Delta t} = -\widehat{\Phi}^* \nabla_s \cdot \mathbf{v}^* - \overline{\Phi} \cdot \left[\omega \nabla_s \cdot \mathbf{v}^+ + (1 - \omega) \nabla_s \cdot \mathbf{v}^- \right]. \tag{8.16}$$

The next step in the discretization process is to apply the tangential divergence operator to (8.16) and insert the result into (8.15), which gives an equation

$$-\Delta_s \widehat{\Phi}^+ + c_H \widehat{\Phi}^+ = RHS. \tag{8.17}$$

This is a Helmholtz-type equation with

$$c_H = \frac{1}{\overline{\Phi}(\omega \Delta t)^2},$$

and RHS is a right hand side composed of known (and extrapolated) terms:

$$RHS = \frac{1 - \omega}{\omega} \Delta_s \Phi^- + c_H \Phi^- +$$
$$+ \Delta_s \Phi_H^+ + \frac{1 - \omega}{\omega} \Delta_s \Phi_H^- -$$
$$- c_H \overline{\Phi} \Delta t \nabla_s \cdot \mathbf{v}^- - c_H \Delta t \Phi^* \nabla_s \cdot \mathbf{v}^* +$$
$$+ \frac{1}{\omega} \nabla_s \cdot \left[f_C (\mathbf{e}_r \times \mathbf{v}) \right]^*.$$

Δ_s is the Laplace-Beltrami operator on the sphere. Thus, we are left with discretizing a Helmholtz-problem on the sphere. This is accomplished by a finite element method with linear basis functions, yielding a second order spatial discretization suitable for the second order semi-Lagrangian time discretization described above.

The grid is maintained by the grid generation library amatos. The initial grid is the truncated icosahedron (see section 3.5). amatos is used to refine uniformly and locally according to a refinement criterion similar to that in PLASMA-P (previous section). Some preliminary validation has been carried out, using the Williamson test cases [415]. The scheme shows good conservation properties and does not need an artificial diminishing viscosity stabilization. The results reported here are personal communication by Thomas Heinze.

Williamson test case 2 consists of a quasi-stationary geostrophically balanced zonal flow (see fig. 8.10). It is not well suited for adaptive refinement, since it does not comprise any local features. However, it is interesting to observe that a well balanced refinement criterion effectively does not react and maintains a static grid for the whole simulation. This, together with time series of the l^2-error of relative vorticity and geopotential height is depicted in

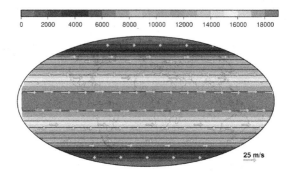

Fig. 8.10. Initial state of the Williamson test case 2.

fig. 8.11. The two different adaptively refined grids (of type CNT[7:11]) that remain almost unchanged during the computation are shown in fig. 8.12.

Test case 5 is well suited for adaptive refinement. An isolated mountain generates Rossby waves in an initially uniform jet-like zonal wind field. Figure 8.13 shows the geopotential height field and some wind vectors after 15 days of simulation. A corresponding grid is depicted in fig. 8.14. Again, similar conservation properties as in test case 2 can be observed.

8.4 Wave Dispersion

This section describes a finite element/finite volume discretization of the shallow water equations in spherical coordinates for wave dispersion (or tsunami modeling) in the ocean. The model accomplishing this task is based on the

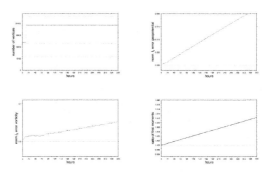

Fig. 8.11. Time series showing the accuracy of PLASMA-FEMmE with Williamson test case 2. (Upper left) number of vertices; (upper right) evolution of the l^2-error of geopotential height; (lower left) evolution of the l^2-error of relative vorticity; (lower right) evolution of the ratio of first moments (mass)

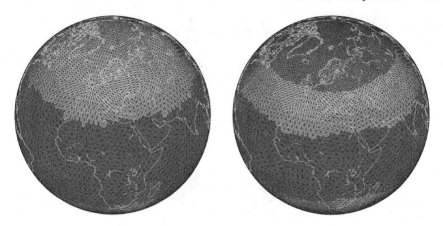

Fig. 8.12. Adaptively refined meshes corresponding to the Williamson test case 2. (**Left**) initial grid; (**right**) grid after first (and only) adaptation step in time step 57. Shadings indicate different grid levels

shallow water equations, given below. In the description, we closely follow the presentation of the non-adaptive but unstructured grid version of the code by Hanert et al. [187] and Danilov [106].

We start with stating the equations for the surface elevation of a fluid (similar to (C.0.13)), the sea surface height (SSH) h and the velocity $\mathbf{v}$ in Eulerian form:

$$\frac{\partial h}{\partial t} + \nabla \cdot [(h + H)\mathbf{v}] = 0, \tag{8.18}$$

$$\frac{\partial \mathbf{v}}{\partial t} + v \cdot \nabla \mathbf{v} + f_C(\mathbf{e}_r \times \mathbf{v}) = -g\nabla h. \tag{8.19}$$

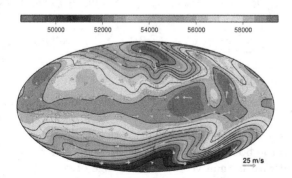

Fig. 8.13. Williamson test case 5 situation after 15 days; geopotential height as isolines/colors, wind as arrows

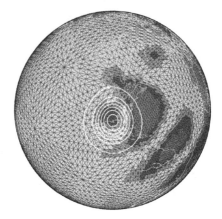

Fig. 8.14. A typical adaptively refined mesh corresponding to Williamson test case 5. Shadings indicate different refinement levels, contour lines represent the hill

Here, H denotes a reference water depth (layer height), and – as usual – f_C, $\mathbf{e}_r$, g, and ∇ are the Coriolis parameter, the outward unit vector on the sphere, the gravitational acceleration constant, and the 2D (spherical) gradient operator, resp.

Since discretization is achieved by a finite element method, we have to derive the variational formulation.

$$\sum_l \int_{\tau_l} \left(\frac{\partial h}{\partial t} b_h - (h + H)\mathbf{v} \cdot \nabla b_h \right) \, d\mathbf{x} +$$

$$+ \sum_l \int_{\partial \tau_l} (h + H)b_h \mathbf{v} \cdot \mathbf{n}_l \, d\Gamma = 0 \quad (8.20)$$

$$\sum_l \int_{\tau_l} \left(\frac{\partial \mathbf{v}}{\partial t} b_\mathbf{v} - (\nabla \cdot (\mathbf{v} b_\mathbf{v})) + f_C(\mathbf{e}_r \times \mathbf{v})b_\mathbf{v} + g\nabla h b_\mathbf{v} \right) \, d\mathbf{x} +$$

$$+ \sum_l \int_{\partial \tau_l} (\mathbf{v}\mathbf{v} \cdot \mathbf{n}_l)b_\mathbf{v} \, d\Gamma + \sum_l \int_{\partial \tau_l} [\mathbf{v}][a(b_\mathbf{v})] \, d\Gamma = 0, \quad (8.21)$$

where we used the following notation:

- $b_\mathbf{v}$, and b_h represent the basis functions corresponding to the variables $\mathbf{v}$, and h.
- $\int_{\tau_l}(\cdot) \, d\mathbf{x}$ is the integral over an element τ_l.
- $\int_{\partial \tau_l}(\cdot) \, d\Gamma$ is the boundary integral over the edges of τ_l.
- $\mathbf{n}_l$ is the outward normal vector at each edge of τ_l.
- $[\cdot]$ represents a jump condition at each edge of a cell τ_l in the following way: let τ_l and τ_k (let $l > k$) be the two adjacent cells of the edge in question, and let g be a constituent. Then

$$[g] = g|_{\tau_l} - g|_{\tau_k}.$$

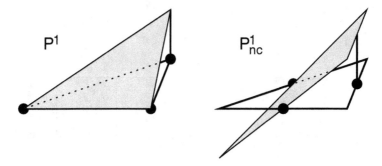

Fig. 8.15. Elements with position of degrees of freedom and basis function examples for the wave dispersion example application; (**left**) P^1 conforming element for sea surface height h, (**right**) P^1_{nc} non-conforming element for velocity **v**

Thus, the last term in (8.21) represents a continuity condition on **v** over edges.

The finite element spaces are chosen in the following way: h (and consequently H) is discretized by a linear conforming Lagrange polynomial P^1 element, while **v** is discretized by a non-conforming linear Lagrange polynomial P^1_{nc} element. The node positions and an example basis function are depicted in fig. 8.15.

Time discretization is realized by a simple, explicit leap-frog (centered difference in time) scheme. Additionally, the equations are augmented by a viscosity term $\nabla\nu\nabla\mathbf{v}$ partly for stabilization, partly because of more realistic results, especially when considering the reflection and propagation of short waves at boundaries. Additionally a bottom drag term $(r + C_d|\mathbf{v}|)\mathbf{v}/H$ is added in order to include realistic effects of tsunami run-up behavior. Here r is a Rayleigh friction coefficient, and C_d is the drag coefficient. So, the discretized equations read

$$\mathbf{v}^+ = \mathbf{v}^- + \Delta t f^-_\mathbf{v}, \tag{8.22}$$
$$h^+ = h^- + \Delta t f^-_h, \tag{8.23}$$

where $f^-_\mathbf{v}$ and f^-_h are all known terms and can be interpreted as flux-terms. Filtering has to be applied in order to prevent splitting modes in the leap-frog scheme.

The adaptive mesh refinement is achieved differently to the approaches in the previous sections. In this method only one grid is employed. Constituents at known times have to be interpolated, once new nodes are inserted. Interpolation is performed at the (generic) order of accuracy provided by the finite element approximation.

The procedure is summarized by the following algorithm:

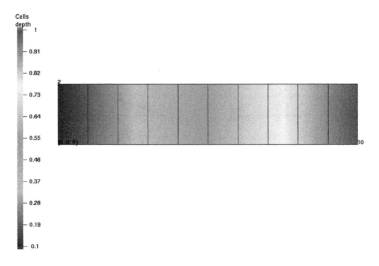

Fig. 8.16. The computational domain for the wave dispersion model. Contours show the depth of the domain

Algorithm 8.4.1 *(Discretization of wave dispersion model)*
Let initial values for **v**, *and h be given, and let* $\mathcal{G}$ *be the computational domain.*

1. *Triangulate the domain do obtain a triangulation* $\overline{\mathcal{T}} = \bigcup_{l=1:M} \overline{\tau_l}$ *with elements* τ_l, $l = 1 : M$, *and edges* γ_k, $k = 1 : L$
2. *Discretize surface height and velocity by* P^1 *conforming and* P^1_{nc} *non-conforming finite element expansions, respectively.*
3. *initialize* h^-, *and* $\mathbf{v}^-$ *by initial values.*
4. *compute* h^+ *and* $\mathbf{v}^+$ *by (8.22) and (8.23).*
5. *Determine a refinement criterion and adapt the mesh correspondingly.*
6. *IF mesh has changed, THEN:*
   a) interpolate h^- *and* $\mathbf{v}^-$ *at the new nodes*
   b) GO TO step 4.

Note that with a fixed time step the algorithm is not unconditionally stable (in contrast to the semi-Lagrangian schemes, described in sections 8.1 and 8.3). Therefore the Courant number is controlled during the computation.

Some preliminary results from this algorithm are given below. The algorithm is applied to a wave dispersion problem in an idealized setting. A rectangular domain and a bottom with a slope form the computational domain (see fig. 8.16).

A radial sine shaped wave form is initialized. The wave initially propagates symmetrically radially. Lateral boundaries reflect the waves and the bottom topography leads to a non-symmetric wave form after some time steps. Figure 8.17 shows this behavior.

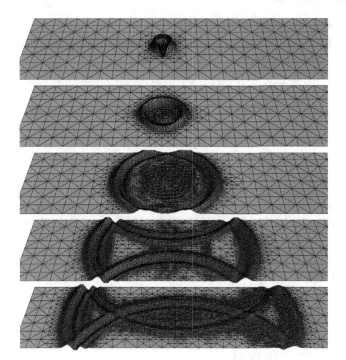

Fig. 8.17. Time series of grids with surface height for the wave dispersion example

8.5 Test Cases

In this section we intend to collect a number of test cases either used in the examples or cited from other publications and especially useful for adaptive computations.

Test Cases for Linear Advection

We start our collection with some test cases for linear advection of a passive tracer in 2D and 3D. These test cases comprise constant divergence-free flow fields as well as divergent fields. They are especially suited for testing conservation properties of advection schemes.

We consider the (conservative) flux form of the advection equation for a mixing ratio or a density ρ:

$$\partial_t \rho + \nabla \cdot (\rho \mathbf{v}) = 0 \quad \text{in } \bar{\mathcal{G}}, \tag{8.24}$$

where the computational domain is defined by $(\mathbf{x}, t) \in \bar{\mathcal{G}} = \mathcal{G} \times I$, $\mathcal{G} \subset \mathbb{R}^d$, $d = 2, 3$, the spatial domain, $I = [0, T]$ the time interval, and $\mathbf{v} = \mathbf{v}(\mathbf{x}, t) \in \mathbb{R}^d$ is a given multi-dimensional flow. In order to be properly defined, (8.24) needs additional specifications for the computational domain $\bar{\mathcal{G}}$, for the boundary conditions $\rho_b(\mathbf{x}, t)$, for the initial condition $\rho_0(\mathbf{x})$, and the wind $\mathbf{v}(\mathbf{x}, t)$. For all 2D tests we will assume

$$\bar{\mathcal{G}} = \bar{\mathcal{G}}_{2D} = \{(\mathbf{x}, t) : \mathbf{x} \in [-0.5, 0.5]^2; \ t \in [0, T]\},$$

where T is specified in the test case. For the 3D tests, we assume

$$\bar{\mathcal{G}} = \bar{\mathcal{G}}_{3D} = \{(\mathbf{x}, t) : \mathbf{x} \in [0, 1]^3; \ t \in [0, T]\}.$$

Additionally, in 3D we will denote with φ_H the horizontal component of a vector-valued entity $\varphi \in \mathbb{R}^3$. Thus, if $\varphi = (\varphi_x, \varphi_y, \varphi_z)^T$ then $\varphi_H = (\varphi_x, \varphi_y)^T$.

We will assume all lateral boundaries to be outflow boundaries. That means, mass can escape from the computational domain, but can not enter. It remains to be specified the initial condition, the wind, and the time interval.

Example 8.5.1 *(Diagonal Wind)*
   The 2D case:

$$\rho_0(\mathbf{x}, t) = \begin{cases} 1, & \text{in } \{\mathbf{x} : |\mathbf{x} - (-0.25, -0.25)^T| < 0.15\}, \\ 0, & \text{else.} \end{cases}$$

$$\mathbf{v}(\mathbf{x}, t) = (1, 1)^T \cdot s, \text{ with a scaling factor } s = 0.36361^{-5}$$

$$T = 18,000 \text{ s}$$

The 3D case:

$$\rho_0(\mathbf{x}, t) = \begin{cases} 1, & \text{in } \{\mathbf{x} : |\mathbf{x} - (0.25, 0.25, 0.5)^T| < 0.15\}, \\ 0, & \text{else.} \end{cases}$$

$$\mathbf{v}_H(\mathbf{x}, t) = (1, 1)^T \cdot s, \text{ with a scaling factor } s = 0.36361^{-5}$$

$$\mathbf{v}_z(\mathbf{x}, t) = 0$$

$$T = 18,000 \text{ s}$$

Example 8.5.2 *(Converging Wind)*
   The 2D case:

$$\rho_0(\mathbf{x}, t) = \begin{cases} 1, & \text{in } \{\mathbf{x} : |\mathbf{x} - (-0.25, 0.0)^T| < 0.15\}, \\ 0, & \text{else.} \end{cases}$$

$$\mathbf{v}(\mathbf{x}, t) = ((0.75, 0)^T - \mathbf{x}^T) \cdot \frac{\omega}{2}, \text{ with } \omega = 0.36361^{-4}$$

$$T = 18,000 \text{ s}$$

The 3D case:

$$\rho_0(\mathbf{x}, t) = \begin{cases} 1, & \text{in } \{\mathbf{x} : |\mathbf{x} - (0.25, 0.5, 0.5)^T| < 0.15\}, \\ 0, & \text{else.} \end{cases}$$

$$\mathbf{v}(\mathbf{x}, t) = ((1.25, 0.5, 0.5)^T - \mathbf{x}^T) \cdot \frac{\omega}{2}, \text{ with } \omega = 0.36361^{-4}$$

$$T = 18,000 \text{ s}$$

The following *slotted cylinder test case* has been adopted from a test case originally proposed in [425]. Note that the angular velocity factor ω is chosen such that 96 time steps of $1,800$ s are exactly one revolution. This test case represents a hard test for the numerical diffusion of an advection scheme, since it defines a non-smooth tracer function. Adaptive grids help to reduce numerical diffusion substantially while maintaining efficiency. Additionally, grid refinement increases the order of approximation near the steep gradients. High order approximations tend to show oscillation effects (over and under shooting) near these "singularities".

Example 8.5.3 *(Slotted Cylinder Test Case)*
The 2D case:

$$\rho_0(\mathbf{x},t) = \begin{cases} 1, \text{ in } \{\mathbf{x} : |\mathbf{x} - (-0.25, 0.0)^T| < 0.15, \\ \quad \text{ and } \mathbf{x} \notin [-0.28, -0.22] \times [0, -0.5]\}, \\ 0, \text{ else.} \end{cases}$$

$$\mathbf{v}(\mathbf{x},t) = ((-y, x)^T \cdot \omega, \text{ with } \omega = 0.36361^{-4}$$

$$T = 172, 800 \text{ s}$$

The 3D case:

$$\rho_0(\mathbf{x},t) = \begin{cases} 1, \text{ in } \{\mathbf{x} : |\mathbf{x} - (0.25, 0.5, 0.5)^T| < 0.15, \\ \quad \text{ and } \mathbf{x} \notin [0.22, 0.28] \times [0.5, 1.0] \times [0, 1.0]\}, \\ 0, \text{ else.} \end{cases}$$

$$\mathbf{v}_H(\mathbf{x},t) = ((-y, x)^T \cdot \omega, \text{ with } \omega = 0.36361^{-4}$$

$$\mathbf{v}_z(\mathbf{x},t) = 0$$

$$T = 172, 800 \text{ s}$$

The *accelerating wind* test case has been adopted and extended to 3D from [218]. Additionally, the scaling has been adjusted such that after 72 time steps of $1,800$ s a full revolution is complete, and analytic error evaluation can be performed. As an example we give snapshots of both the 2D and 3D simulation of the accelerated wind field test case in figs. 8.18 and 8.19 resp.

Example 8.5.4 *(Accelerating Wind)*
The 2D case:

$$\rho_0(\mathbf{x},t) = \begin{cases} 1, \text{ in } \{\mathbf{x} : |\mathbf{x} - (-0.25, 0.0)^T| < 0.15, \\ \quad \text{ and } \mathbf{x} \notin [-0.28, -0.22] \times [0, -0.5]\}, \\ 0, \text{ else.} \end{cases}$$

$$\mathbf{v}(\mathbf{x},t) = \begin{cases} (-y, x)^T \cdot \omega, & \text{if } = (x, y) \in \{\mathbf{x} : x \le 0\}, \\ (-y, x)^T \cdot \omega(1.5 \cos(2\phi) + 2.5), & \text{if } = (x, y) \in \{\mathbf{x} : x > 0, y > 0\}, \\ (-y, x)^T \cdot \omega(1.5 \cos(2\psi) + 2.5), & \text{if } = (x, y) \in \{\mathbf{x} : x > 0, y \le 0\}, \end{cases}$$

$$\text{with } \omega = 0.36361^{-4}, \quad \phi = \arctan(\frac{y}{x}), \quad \psi = \arctan(\frac{-y}{x}).$$

$$T = 129, 600 \text{ s}$$

Fig. 8.18. Accelerating wind test case in 2D: Initial field of tracer distribution (**left**) intermediate state after 36 time steps (**center**) and final step after 72 time steps (**right**)

The 3D case:

$$\rho_0(\mathbf{x}, t) = \begin{cases} 1, \text{ in } \{\mathbf{x} : |\mathbf{x} - (0.25, 0.5, 0.5)^T| < 0.15, \\ \quad \text{and } \mathbf{x} \notin [0.22, 0.28] \times [0.5, 1.0] \times [0, 1.0]\}, \\ 0, \text{ else.} \end{cases}$$

$$\mathbf{v}_H(\mathbf{x}, t) = \begin{cases} (-y, x)^T \cdot \omega, & \text{if } = (x, y) \in \{\mathbf{x} : x \le 0\}, \\ (-y, x)^T \cdot \omega(1.5\cos(2\phi) + 2.5), & \text{if } = (x, y) \in \{\mathbf{x} : x > 0, y > 0\}, \\ (-y, x)^T \cdot \omega(1.5\cos(2\psi) + 2.5), & \text{if } = (x, y) \in \{\mathbf{x} : x > 0, y \le 0\}, \end{cases}$$

$$\text{with } \omega = 0.36361^{-4}, \quad \phi = \arctan(\frac{y}{x}), \quad \psi = \arctan(\frac{-y}{x}).$$

$$\mathbf{v}_z(\mathbf{x}, t) = 0$$

$$T = 129,600 \text{ s}$$

Note that since we deal with linear advection, all the previous test cases can be solved analytically and, therefore, are very well suited for accuracy investigation. However, convergence tests need a smooth density distribution function, since even high order schemes usually do not achieve high order convergence near discontinuities in the data. Therefore, a fifth test case is given

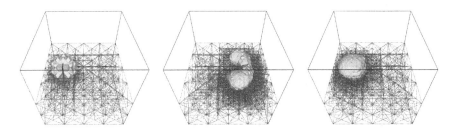

Fig. 8.19. Accelerating wind test case in 3D: Initial field of tracer distribution (**left**) intermediate state after 36 time steps (**center**) and final step after 72 time steps (**right**). The 0.95 iso-surface is shown

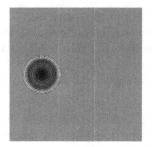

Fig. 8.20. Initial field of tracer distribution for convergence test case, contour lines in 0.1 unit steps

for convergence tests. We rotate the initial situation by just $1/4$ revolution. The initial configuration is depicted in fig. 8.20.

Example 8.5.5 *(Convergence test case)*
  The 2D case:

$$\rho_0(\mathbf{x}, t) = \begin{cases} 4\cos\left(\frac{r\pi}{2R}\right), & \text{in } \{\mathbf{x} : r < R\} \\ & \text{with } r = \|\mathbf{x} - (-0.25, 0.0)^T\|, \ R = 0.15; \\ 0, & \text{else.} \end{cases}$$

$$\mathbf{v}(\mathbf{x}, t) = (-y, x)^T \cdot \omega, \text{ with } \omega = 0.36361^{-4}$$
$$T = 43,200 \text{ s}$$

The 3D case:

$$\rho_0(\mathbf{x}, t) = \begin{cases} \cos\left(\frac{r\pi}{2R}\right), & \text{in } \{\mathbf{x} : r < R\} \\ & \text{with } r = \|\mathbf{x} - (0.25, 0.5, 0.5)^T\|, \ R = 0.15; \\ 0, & \text{else.} \end{cases}$$

$$\mathbf{v}_H(\mathbf{x}, t) = (-y, x)^T \cdot \omega, \text{ with } \omega = 0.36361^{-4}$$
$$\mathbf{v}_z(\mathbf{x}, t) = 0$$
$$T = 43,200 \text{ s}$$

Test Cases for the Shallow Water Equations

The shallow water equations serve as a first mile stone in the development of dynamical cores for atmospheric modeling systems. Most authors use the well known *Williamson test cases*, published in [415] with a set of reference solutions [222]. We are not going to repeat these test cases here.

One weakness of the Williamson test cases is that there is no time-dependent dynamical test case with an analytical solution. One such test case was proposed by Läuter, Handorf, and Dethloff recently (hereafter referred to as [LHD05]) [257]. Test cases for the spherical shallow water equations are grouped into four categories:

- analytical solutions to simplified equations;
- steady state solutions to the complete equations;
- unsteady analytical solutions to the full equations with a prescribed solution forcing technique;
- close-to-reality initial data and comparison to high resolution model reference solutions.

[LHD05] add a new class of analytical test cases that does not need a forcing technique, but should converge under correct numerical treatment of the continuous equations. The derivation of test cases is based on the transformation of known solutions to the shallow water equations in non-rotating frames of reference to the rotating sphere. We cite two of the given test cases here, omitting the technique of transformation, which is described in detail in [LHD05] and has been adopted from [134, 311].

We start with repeating the formulation of the shallow water equations in a spherical rotating frame.

$$\frac{d\mathbf{v}}{dt} + \nabla_s \Phi + f_C \mathbf{n} \times \mathbf{v} + \frac{|\mathbf{v}|^2}{|\mathbf{x}|}\mathbf{n} = 0,$$

$$\frac{d\Phi}{dt} + \Phi \nabla_s \cdot \mathbf{v} - \nabla_s \cdot (\mathbf{v}\Phi_0) = 0, \tag{8.25}$$

$$\mathbf{v} \cdot \mathbf{n} = 0.$$

As usual, $\mathbf{v}$ denotes the tangential velocity, $\mathbf{n}$ the unit outward normal vector, f_C the Coriolis term, Φ the geopotential height and Φ_0 the orography. A simple test case is given by the axially symmetric flow:

Example 8.5.6 *(Unsteady axially symmetric flow)*
We first define

- $\mathbf{a} \in \mathbb{R}^3$, $|\mathbf{a}| = 1$ *a fixed symmetry axis, and k_1, k_2 constants;*
- $u : [-1, 1] \to \mathbb{R}$ *a smooth integrable velocity profile, i.e. u satisfies*

$$\int_{[-\frac{\pi}{2}, \frac{\pi}{2}]} |\tan x u^2(\sin x)| \, dx < \infty.$$

- φ_t *a coordinate transformation function from the non-rotating to the rotating frameset given by*

$$\varphi_t(\mathbf{x}) = (\mathbf{x} \cdot \mathbf{b}_i)\mathbf{e}_i, \varphi_t^{-1}(\mathbf{y}) = (\mathbf{y} \cdot \mathbf{e}_i)\mathbf{b}_i,$$

where

$$\mathbf{b}_1(t) = \cos(\Omega t)\mathbf{e}_1 + \sin(\Omega t)\mathbf{e}_2, \ \mathbf{b}_2(t) = -\sin(\Omega t)\mathbf{e}_1 + \cos(\Omega t)\mathbf{e}_2, \ \mathbf{b}_3(t) = \mathbf{e}_3,$$

with Ω the earths rotational velocity.

Then the pair $(\mathbf{v}_A, \Phi_A)$ is a solution to the shallow water equations (8.25) with

$$\mathbf{v}_A(\mathbf{x}, t) = u(\varphi_t(\mathbf{a}) \cdot \mathbf{n}(\mathbf{x})) \frac{\varphi_t(\mathbf{a}) \times \mathbf{n}(\mathbf{x})}{|\varphi_t(\mathbf{a}) \times \mathbf{n}(\mathbf{x})|} - \boldsymbol{\Omega} \times \mathbf{x},$$

$$\Phi_A(\mathbf{x}, t) = \int_0^{\arcsin(\varphi_t(\mathbf{a}) \cdot \mathbf{n}(\mathbf{x}))} \tan y \, u^2(\sin y) \, dy + \frac{1}{2}(\boldsymbol{\Omega} \cdot \mathbf{x}) + k_1,$$

$$\Phi_{A,0}(\mathbf{x}) = \frac{1}{2}(\boldsymbol{\Omega} \cdot \mathbf{x}) + k_1.$$

We used $\boldsymbol{\Omega} = \Omega \mathbf{e}_3$, and $\Phi_{A,0}$ denotes the orography. Note that k_1, and k_2 are not specified and can be chosen arbitrarily.

Remark 8.5.7 *With the analytical solution given in test case 8.5.6, the corresponding analytical functions for the vorticity ζ and the divergence δ can be given (r denotes the sphere's radius):*

$$\zeta(\mathbf{x}, t) = \frac{u(\varphi_t(\mathbf{a}) \cdot \mathbf{n}(\mathbf{x})) \varphi_t(\mathbf{a}) \cdot \mathbf{n}(\mathbf{x})}{r \sqrt{1 - (\varphi_t(\mathbf{a}) \cdot \mathbf{n}(\mathbf{x}))^2}}$$

$$- \frac{1}{r} u'(\varphi_t(\mathbf{a}) \cdot \mathbf{n}(\mathbf{x})) \sqrt{1 - (u'(\varphi_t(\mathbf{a}) \cdot \mathbf{n}(\mathbf{x})))^2} - 2\boldsymbol{\Omega} \cdot \mathbf{n}(\mathbf{x}),$$

$$\delta(\mathbf{x}, t) = 0.$$

The second test case cited from [LDH05] is a jet stream. A sketch of that test case is given in fig. 8.21.

Example 8.5.8 *(Unsteady jet stream)*
   We start with some definitions:

- *Let $\mathbf{c} \in \mathbb{R}^3$ with $|\mathbf{c}| = 1$ a fixed rotation axis.*
- *Let $u_0, u_1 \in \mathbb{R}$ flow velocities, $\theta_0, \theta_1 \in [-\frac{\pi}{2}, \frac{\pi}{2}]$ latitudinal angles, and $k_1, k_2 \in \mathbb{R}$ constants.*
- *Define the velocity and geopotential of a solid body rotation $\mathbf{v}_R$ and Φ_R as*

$$\mathbf{v}_R(\mathbf{x}, t) = u_0 \varphi_t(\mathbf{c}) \times \mathbf{n},$$

$$\Phi_R(\mathbf{x}, t) = \frac{1}{2} \left[-(u_0 \varphi_t(\mathbf{c}) \cdot \mathbf{n} + \boldsymbol{\Omega} \cdot \mathbf{x})^2 + (\boldsymbol{\Omega} \cdot \mathbf{x})^2 \right] + k_1.$$

- *With $c_0 = u_1 \exp(4(\theta_0 - \theta_1)^{-2})$, let the jet velocity profile u_J be*

$$u_J(x) = \begin{cases} c_0 \exp\left((\arcsin(x - \theta_0) \arcsin(x - \theta_1))^{-1}\right) & \text{for } \theta_0 \leq \arcsin(x) \leq \theta_1, \\ 0 & \text{else.} \end{cases}$$

- *Let $\mathbf{a} = \frac{u_0 \mathbf{c} + r\boldsymbol{\Omega}}{|u_0 \mathbf{c} + r\boldsymbol{\Omega}|}$, r the sphere's radius, and a second velocity profile defined by*

$$u_S(x) = (u_0 \mathbf{c} + r\boldsymbol{\Omega}) \sqrt{1 - x^2}$$

for $x \in [-1, 1]$.

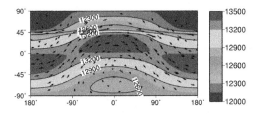

Fig. 8.21. Velocity vectors and geopotential height contours for test case 8.5.8

With these pre-definitions and the notation $s = \sin(y)$ the following system is a solution to the shallow water equations (8.25):

$$\mathbf{v}(\mathbf{x}, t) = \mathbf{v}_R(\mathbf{x}, t) + u_J(\varphi_t(\mathbf{a}) \cdot \mathbf{n}) \frac{\varphi_t(\mathbf{a}) \times \mathbf{n}}{|\varphi_t(\mathbf{a}) \times \mathbf{n}|},$$

$$\Phi(\mathbf{x}, t) = \Phi_R(\mathbf{x}, t) - \int_0^{\arcsin(\varphi_t(\mathbf{a}) \cdot \mathbf{n})} \tan(y) \left(u_S(s) u_J(s) \right) + u_J^2(s) \right) dy + k_1 \quad (8.26)$$

$$\Phi_0(\mathbf{x}) = \frac{1}{2} (\boldsymbol{\Omega} \cdot \mathbf{x})^2 + k_2. \quad (8.27)$$

Note that with the definition of an analytical solution in test case 8.5.8 again vorticity and divergence can also be given. Indeed, they are identical to those in test case 8.5.6. Note further that [LDH05] gives also solutions to the shallow water equations, formulated in spherical coordinates for convenience. The constants used in the test cases are

k_1: is the minimum geopotential height and should be chosen large enough to impose suitable layer of air above orography. In [LDH05] this is $k_1 = 2000$.

k_2: is the bottom geopotential height and usually set to $k_2 = 0$.

u_0: is the mean (westerly) velocity and in [LDH05] set to $u_0 = 20 [ms^{-1}]$.

u_1: is the maximum velocity in the jet and set to $u_1 = 38 [ms^{-1}]$ in accordance with the Williamson test cases.

θ_0: is the southern latitude bounding the jet and set to $\theta_0 = \frac{\pi}{7}$.

θ_1: is the northern latitude bounding the jet and set to $\theta_1 = \frac{\pi}{2} - \theta_0$.

We close this section with a list of common test cases for the shallow water equations.

- Galewsky and coworkers propose a barotropic instability test case in [159].
- We mentioned before the well known Williamson test suite of seven test cases including solid body rotation, quasi stationary waves (Rossby-Haurwitz waves), instationary flow and near realistic initial value problems with reference solutions [415, 222].
- A balanced initial state of wave number 1, proposed by McDonald and Bates [282].
- Solid body rotation test cases with an inclination angle [398, 414].
- Zonal wind steady state solution by Browning and coworkers [74].

As an outlook, we mention the three dimensional test cases for baroclinic dynamical cores of future numerical modeling systems, proposed by Jablonowski and by Polvani and coworkers. These test cases are especially interesting, since they provide a frame of reference for full 3D models. The Jablonowski-Williamson test case describes the evolution of a baroclinic wave from a balanced (and perturbed) initial state [221]. It is well suited for adaptive methods, since the wave stays spatially local. A somewhat similar test case, also starting from an initially balanced flow field that is perturbed, has been proposed by Polvani and coworkers for primitive equation solvers [324]. Earlier test cases for 3D dynamical cores have been proposed by Held and Suarez [195] and Boer and Denis [64].

For fully featured models, the only way to evaluate their quality is by comparing modeling results to measurements. For an adaptive global atmospheric model such an evaluation is documented by Boybeyi and coauthors [69].

9

Conclusions

Adaptive atmospheric modeling is a truly interdisciplinary approach in scientific computing. We have seen several tools from applied mathematics, theoretical physics, computer science and even pure mathematics interacting in order to achieve efficiency, accuracy and robustness for adaptive algorithms in atmospheric modeling. This chapter aims in reviewing the methods and evaluating their respective usefulness. From this we try to derive a path for future research directions.

9.1 Evaluation of Methods

It is indeed challenging and fascinating at the same time how broad the spectrum of mathematical methods that help solving adaptive atmospheric modeling. We touched

- graph theory and computational geometry in the mesh generation part and parallel mesh partitioning;
- functional analysis with Sobolev spaces in the finite element discretization of differential equations;
- variational formulations of conservation laws;
- geometric and quadrature integration in the discretization of conservative semi-Lagrangian advection;
- classical analysis and set theory in the derivation of space-filling curves;
- complexity theory when deriving optimal search strategies and parallel algorithms.

Above all is the aim of applying adaptive multi-scale methods for true life applications.

In essence, adaptive methods have gained a technical state of maturity that allows for serious applications in atmospheric modeling. Efficient data structures for handling adaptively refined meshes are available and prove to be useful. When applying adaptive mesh refinement, it is important to adhere to two principles

- hierarchical data structures, and
- neighborhood preservation.

There are powerful refinement and de-refinement strategies available, either for quadrilateral/hexahedral refinement and for triangular/tetrahedral refinement. The coupling with automatic mesh generation, however, has not yet been integrated. Most mesh handling and generating tools start from a given coarse grid.

Parallelization, though much more involved with adaptive methods, is possible and practical. Many tools show efficiency up to at least moderate numbers of processors. It is important to consider

- dynamic,
- parallelizable, and
- locality preserving

load balancing strategies.

The construction of differential operators for irregular mesh structures reveales some problematic issues. Mesh independency is not yet fully achieved. There remains some more research to be conducted in order to find a consistent way to deal with mesh irregularities. This is of great importance, since gradients in many discretizations of 3D atmospheric and oceanic circulation play a crucial role for the stability of the whole dynamical core.

The development of discretization schemes for conservation laws with adaptive mesh refinement has gained a state of maturity over the past 15 years that makes complex simulations feasible. We introduced finite volume schemes, discontinuous Galerkin schemes and conservative semi-Lagrangian schemes for solving the dynamic equations of atmospheric circulation. For adaptive methods to be useful, discretization techniques for conservation laws must be capable of implementing

- unstructured
- direction independent

discrete evolution operators. While the first two methods (FV and DG) are well established and proved their effective conservation properties, they lack the advantageous stability property making them truly mesh independent. Semi-Lagrangian schemes on the other hand are stable and therefore more mesh independent, but it is more cumbersome to achieve conservation.

The examples demonstrated in chap. 8 prove the effectiveness of adaptive methods in atmospheric modeling. However, it remains to be shown that adaptivity yields the same advantage in fully complex atmospheric applications. Some test cases, tailored for adaptive modeling, have been proposed. Yet, more interesting and challenging cases are to be developed. Adaptive methods can reveal their effectiveness in cases with

- localized features,
- large scale differences,
- time dependent local phenomena.

This work has tried to give a fair overview of the state of the art of adaptive atmospheric modeling. Some impressive results have been achieved. The interesting work has just begun. Now, we can start of thinking to solve problems that could not be tackled with the methods available so far. Studying fine-scale interaction with global processes, resolving orography realistically, including fully dynamic convective processes, to name a few, appear to be feasible options now.

9.2 Road Map for the Next Five Years

After more than 15 years from first papers on adaptive atmospheric modeling [368, 369, 125], this overview of the current state of adaptive methods for atmospheric modeling reveals: It is by far not a completed field of research, in contrary the real work just started. There are important tasks ahead of the community of adaptive atmospheric modelers and this section tries to outline a road map for the next five years. The reader is asked to be forbearing with the author, since even with some experience in adaptive modeling, there is no looking glass into the future. However, several options for further development are obvious.

Consistent Numerical Methods

There is still a great need for *consistent numerical methods*. That means the quest for the ultimate numerical scheme that is mesh independent, efficient, stable, accurate and preserves the structural properties of the continuous problem, is still open.

We came much closer to this goal in the recent years. Discontinuous Galerkin methods promise to preserve most of the continuous properties of the underlying conservation laws even for high order accuracy. However, there is still some work required to find a stable time integration scheme that fits well to the DG scheme and is stable for large Courant numbers. On the other hand, several authors presented (stable) semi-Lagrangian schemes that were modified such that conservation properties hold [360, 419, 421].

Moreover, it is necessary that the numerical schemes used are robust enough to handle the case of model changes. That means, when refining locally, the assumptions on approximations become more and more violated. In this case, the mathematical model has to be changed or in some cases, the characteristics of the model equations change (e.g. the low mach number flow, where some intermediate state between hyperbolic and parabolic equations is covered). The numerical method must be chosen such that no numerical artifacts result from this fact.

And finally, the numerical method should be able to hide grid effects even in large to very large ratios of refinement to de-refinement mesh size. In order to exploit adaptive methods, one wishes to be able to refine locally down to fine scales, while a global coarse background model generates the necessary background data. The numerical scheme should be stable for this kind of situations.

Refinement Criteria

Adaptive methods are only as good as the refinement criteria driving the adaptation strategy. This fact cannot be over exaggerated. So far, refinement is controlled by

- error proxies;
- physical criteria, like gradients, divergence or vorticity values;
- mathematical error estimates based on discretization or truncation error estimates for the dynamic equations.

However, what if there are several constituents that all play a role in the accuracy of the simulated situation, but react differently in distinct regions of the computational domain? How to treat equations, where the error can propagate in time? How to detect sub-grid scale processes that may trigger a refinement once they are relevant in magnitude (but cannot be resolved on a coarse mesh)?

Since atmospheric modeling is always concerned with multi-scale interaction of processes with a multitude of constituents involved, there is not much research from other communities that could serve as template for adoption. Research efforts will be necessary to solve the problem of finding rigorous efficient and robust refinement criteria.

Sub-Grid Parametrization

As mentioned above, atmospheric modeling is concerned with multi-scale phenomena, spanning a range of scales that cannot be resolved even with adaptive grid refinement. Therefore, some processes will always be captured by intelligent parametrization techniques. However, current parametrization is most often tuned to certain mesh resolutions. Thus, a re-design of parametrization schemes such that they become grid-independent or can be adapted to changing grid sizes in a functional way, is necessary.

This effort, as painful as it might be, promises some fruitful insight. Since such mesh-independent parametrization will probably be derived from first principles, a more consistent and physical approach may eventually be the result.

It is - on the other hand - naive to assume that after so many years of intense research activity in parameterizations that the community will easily take the challenge of rethinking many of the methods used so far. Therefore, this research issue will probably remain on the agenda far beyond the next five years.

Additionally, the evolution of promising stochastic approaches to sub-grid parametrization are not prepared to adaptive mesh sizes. It may well be that stochastic modeling is much more mesh independent than the physics based parametrization, but this has not yet been investigated.

Computational Efficiency

Since adaptive methods almost always involve some kind of indirect addressing, irregularity in data structures and changing memory requirements, there is still a great demand in more efficient computational methods. Additionally, the first adaptive atmospheric models have to catch up with the vast optimization efforts that went into gaining speed in operational models and corresponding numerical cores. We outlined several methods in this work that try to address this issue, however, it still remains to be shown that adaptive methods really outperform well tuned uniform and regular grid based models.

Realistic and Useful Applications

Up to now, the state of the art in adaptive atmospheric modeling allowed for the development of dynamical cores and simplified models only. The only operational adaptive model OMEGA may be considered an exception, however, it also does not exploit the full potential of adaptive methods.

I can see applications in severe weather forecasting, where fronts of thunderstorms can be resolved accurately (without parameterizing the local convection). More generally, it is not hard to foresee that meso-scale convection cells can be detected by a sub-grid criterion and then adaptively resolved in a non-hydrostatic model.

For realistic assessment of orographic influence especially near mountainous regions (Föhn wind, lee waves, etc.) a locally refined mesh is an obvious option. Even in large scale models, this kind of refinement might yield much more realistic flow fields than a parameterized orographic forcing.

There are applications of adaptive methods for air quality and pollution modeling of localized emitters. However, in the inverse modeling situation, where one uses an adjoint model for solving the problem of determining a source from field values, adaptive methods can greatly increase the local accuracy, as shown in sect. 8.2. It remains to be shown, that this approach remains useful and applicable for realistic situations with reacting constituents, etc.

Scale-interaction is another promising field for adaptive methods. Large scale processes triggering small scale features that again have an important influence/feed-back to the large scale or small scale processes triggering large scale characteristic features play an important role in the atmosphere and oceans. For example, the tide induced small scale jets that are caused by coastal topography in the Weddel sea are supposedly responsible for significant interaction and deep sea water formation in the Weddel eddy [315]. While the tide wave is a truly global process with a characteristic length scale of $\frac{1}{2}r_{\mathrm{Earth}}$ (r_{Earth} the earth's radius), the jets have a diameter of some hundreds of meters.

A realistic tsunami forecasting model will face the same ratio in magnitudes: The wave caused by deep sea seismic activity and the corresponding characteristic length scale is probably a large fraction of an ocean basin or several thousand kilometers. On the other hand the Tsunami unfolds it's power due to shallow water and an amplification of energy due to convergence. In these regions, characteristic length scales of a few meters play a role.

Rigorous Scale Analysis

In order to be usefully solved by adaptive modeling techniques, it is necessary to know the scale regions for which certain approximations are valid. Furthermore, one can think of methods that extend the equations once they refine below a certain level. In order to control the refinement criterion sensibly, a rigorous scale analysis has to be conducted for the common mathematical models of atmospheric dynamics [274].

Coupling of Sub-System Models

Adaptive grids pose additional difficulties for coupling several models. While regular grids allow for more or less straight forward interpolation techniques that have to adhere to conservation and structural properties of the underlying equations, it is less obvious how to handle this problem with adaptively refined, temporally changing grids. Interpolation is always an option, however, the error has to be controlled.

More generally, what is the interface to an adaptive model? Up to now, atmospheric modelers think in grid structures. This is not a useful concept for adaptive methods. Here, we want to think in functional values, regardless of the mesh. However, if a mesh is no longer explicitly provided, error bounds need to be given together with interchanged values, since otherwise no useful assessment of the overall accuracy is possible.

Pre- and Post-Processing

Most of the tools in atmospheric modeling that can be used for visualization, archiving, data-base storage, distribution and interchange of data, etc., are made for uniform and quadrilateral meshes. A first approach to reconnect adaptive methods to the toolbox available for pre- and post-processing is interpolation. In many cases this will suffice to yield satisfactory results (e.g. in visualization, where the visual effect can be maintained even if the error of interpolation is in the order of a few percent). However, if accurate data is required, an interpolation might not be the right option. It will be necessary to re-design many of the pre- and post-processing tools for irregularly structured data.

Again, this is not necessarily a major drawback, since it might be worth to re-design several of the aged methods anyway. However, it is a time factor that has to be taken into account.

Vision

It is the author's vision that in some years, we will not think in grids anymore. We will be talking about results, error measures, experiments. The numerical methods will be able to resolve the model equations accurately up to a user supplied tolerance. Results will be handled mesh-independently, more generally meshes will be what they ought to be, just tools for representing data. We will come back to the type or research in the early days of atmospheric sciences, that is we will be concerned about formulating model equations and well posed initial and boundary conditions instead of being concerned about the correct numerical solution of such equations. Adaptive methods will have gained a state of robustness and efficiency that they can be used in modular experimental software environments, allowing for interactive playful research.

A

Some Basic Mathematical Tools

This appendix contains a brief section on basic mathematical tools that are used throughout this book. We denote by $\mathbf{x} = \mathbf{x}_{\mathbf{x}_0,t_0}(t)$ the *position* at time t of a fluid particle that started at position $\mathbf{x}_0$ at time t_0. We will often omit the subscripts. Let $\check{\mathbf{x}} = \check{\mathbf{x}}_{\mathbf{x}_0,t_0}(t)$ denote the *trajectory* representing the path of a particle that starts at position $\mathbf{x}_0$ at starting time t_0 up to time $t > t_0$. In fact $\check{\mathbf{x}}$ marks the path between $\mathbf{x}_0$ and $\mathbf{x}$. $V(t)$ shall denote a *material volume* moving with the fluid and containing exactly the same particles over time.

Obviously, the derivative in time of a position in space is given by the velocity

$$\frac{d\mathbf{x}(t)}{dt} = \mathbf{v}(\mathbf{x},t).$$

The time derivative of a particle's property ψ (the *material property*) is denoted by $\frac{d\psi}{dt}$, called the *material derivative*. Since we assume a dense distribution of particles and the property ψ being defined for each particle, we obtain a scalar field $\psi(\mathbf{x},t)$. For the material derivative, applying the chain rule and taking into consideration that the material derivative tracks individual particles, we have

$$\frac{d\psi}{dt} = \frac{\partial\psi}{\partial t} + \mathbf{v}\nabla\psi. \tag{A.1}$$

We will use the following theorems, that are recalled here for the reader's convenience. Proofs can be found in [410].

Theorem A.0.1 *(Divergence Theorem)*
For any volume $V \subset \mathbb{R}^d$ with piecewise smooth closed surface S and any differentiable vector field $\mathbf{u}$ the following integral relation holds:

$$\int_V \nabla \cdot \mathbf{u} \, dx = \int_S \mathbf{u} \cdot \mathbf{n} \, ds,$$

where dx is the spatial differential, ds is the surface differential and $\mathbf{n}$ is the outward unit normal on S.

Theorem A.0.2 *(Stokes theorem)*
Let C be a closed curve and S a piecewise smooth surface bounded by C. Then, for any differentiable vector field $\mathbf{u}$

$$\oint_C \mathbf{u} \cdot \mathbf{dx} = \int_S (\nabla \times \mathbf{u}) \cdot \mathbf{n} \; ds,$$

where $\mathbf{dx}$ is the curve differential along C, and $\mathbf{n}$ as above the outward unit normal (right handed in traversal direction).

With Stoke's theorem one has a representation of $\mathbf{u}$ in terms of a gradient:

Corollary A.0.3 *(Potential)*
If $\mathbf{u}$ is irrotational, i.e. $\mathbf{u}$ satisfies $\nabla \times \mathbf{u} = 0$, then there exists a scalar field ψ such that
$$\mathbf{u} = \nabla \psi.$$
ψ is called the potential.

Another representation of a vector field $\mathbf{u}$ is given by the

Theorem A.0.4 *(Helmholtz representation)*
For any bounded continuous vector field $\mathbf{u}$ that vanishes at infinity there exists a scalar field ψ and a divergence free vector field $\mathbf{w}$ (i.e. $\mathbf{w}$ satisfies $\nabla \cdot \mathbf{w} = 0$), such that
$$\mathbf{u} = \nabla \psi + \nabla \times \mathbf{w}.$$

The following theorem is used in section 7.4.

Theorem A.0.5 *(Transport theorem)*
For any material volume $V(t)$ and differential scalar field ψ we have ($\mathbf{v}$ the velocity):

$$\frac{d}{dt} \int_{V(t)} \psi \; dx = \int_{V(t)} \left(\frac{\partial \psi}{\partial t} + \nabla \cdot (\mathbf{v}\psi) \right) \; dx. \tag{A.2}$$

B

Metrics for Parallelizing Irregularly Structured Problems

Naturally, all the metrics for measuring the parallel performance of a non-adaptive program apply similarly to an adaptive program. For parallel metrics one may consult the standard literature (e.g. [205]). We recall the most important performance metrics for use in this book:

Definition B.0.6 *(Standard Metrics)*

1. *The computing* **time** *for a single process is denoted by t_p, where p is the process id. We omit the dependence of $t_p = t_p(N)$ on the problem size N unless otherwise stated.*

2. *The* **wall clock time** *t_{total} for a parallel process is defined by the maximum time among each individual process times[1]:*

$$t_{\text{total}} := t_P := \max_{p=1:P} t_p.$$

3. *The* **speedup** *s_P for P processors is defined as the inverse factor of decrease in computing time with parallel processors:*

$$s_P := \frac{t_1}{t_P},$$

where t_1 and t_P is the wall clock time on 1 and P processors respectively.

4. *The parallel* **efficiency** *E_P for P processors is defined as the percentage of real achieved speedup over theoretically optimal speedup. In other words, the efficiency is the speedup scaled by $\frac{1}{P}$.*

$$E_P := \frac{t_1}{P \cdot t_P}.$$

---

[1] Note that we assume that all processes start at the same time. More generally, the wall clock time is the time from program start to program termination with parallel execution in between these two points.

5. The **scale-up** S_l (l scaling factor) measures the behavior of an algorithm with increasing number of processors P and similarly increasing problem size N, such that each process maintains a constant problem size. The main influence limiting the scale-up is inter-process communication.

$$S_l := \frac{t_P(N)}{t_{l \cdot P}(l \cdot N)}.$$

In order to measure the quality of partitions we need to establish metrics for the usefulness of a partitioning strategy. As mentioned in chapter 5, we need to fulfill several requirements. First, a partitioning strategy needs to be efficient. Since partitioning is not our main task, all the work that goes into calculating an (almost) optimal partitioning is overhead. We will account for this overhead. A good partitioner will cause as little overhead as possible.

Definition B.0.7 (Partitioning Overhead)
The partitioning overhead O_{part} is the ratio of time for computing the partitioning t_{part} over total time t_{total}, and is a number $0 < O_p < 1$.

$$O_{\text{part}} := \frac{t_{\text{part}}}{t_{\text{total}}}, \tag{B.1}$$

Remark B.0.8 Note that O_{part} can be a value that depends on the number of processors. Since the partitioning might scale with another slope as the overall program.

Since the overall performance is utterly dependent on good load balancing (note that the parallel wall clock time can only be as short as the largest individual process), we define the load balancing parameter:

Definition B.0.9 (Load Balancing Parameter)
Let N_p be the problem size of processor p. The load balancing parameter is defined by

$$\mathcal{L} := \frac{\max_p N_p}{\min_p N_p}$$

Note that $\mathcal{L} \geq 1$ and that optimal load balancing is achieved when $\mathcal{L} = 1$.

But not only load balancing is crucial for efficient parallelization, data communication is also to be minimized. If inter-process communication has to take place, then we assume that it takes place over interfaces that have distinct processes on either side. Therefore, we introduce two metrics for evaluating the partitioning with communication issues in mind.

Definition B.0.10 (Edge Cut)
The edge cut $\mathcal{E}$ is defined as the number of edges (2D) or faces (3D) resp. $f_{p_i}^{p_j}$ that form a boundary between processors relative to the total number of edges/faces f_{total}:

$$\mathcal{E} := \frac{\# f_{p_i}^{p_j}}{\# f_{\text{total}}}$$

Finally, since re-partitioning has to take place frequently, we also have to consider the cost for moving cells between processors.

Definition B.0.11 *(Re-Partitioning Cost)*
Let M be the total number of mesh cells and let M_i^j be the number of cells that are moved from processor p_i to processor p_j. Then the re-partition cost $\mathcal{R}$ is defined by

$$\mathcal{R} := \frac{\sum_{i,j=1}^{P} M_i^j}{M}.$$

Note that $0 \leq \mathcal{R} \leq 1$, because every cell can be moved at most once (the net effect is counted). If the unknowns are associated to processors (instead of the cells) then the definition of $\mathcal{R}$ has to be modified accordingly.

Some interesting comments on load balancing and metrics concerned with mesh partitioning can be found in [197, 198]. It is noteworthy that the edge cut is often a misleading measure for the quality of a partitioning method, since the computation is influenced by the true communication cost which is only weakly related to the edge cut.

C

Rotating Shallow Water Equations in Spherical Geometries

In sect. 7.1 we derived the shallow water equations from basic conservation principles. Here we extend the equations to be valid on the rotating sphere. Not all applications in atmospheric modeling need to account for the rotation, but since our emphasis is on global modeling, we have to consider it.

We consider a rotating frame of reference with angular velocity Ω. An observer in an inertial frame of reference, which is the frame for the derived equations so far, and a second observer in a rotating frame of reference will have different views of changes to a vector $\xi = (\xi_1, \xi_2, \xi_3)$ in the following way:

$$\left[\frac{d\xi}{dt}\right]_I = \left[\frac{d\xi}{dt}\right]_R + \Omega \times \xi.$$

The subscripts I and R refer to the inertial and rotating frames resp. Using this relation, we see immediately that for the velocity $\mathbf{v}$ we have that $\mathbf{v}_I = \mathbf{v}_R + \Omega \times \mathbf{x}$. Some elementary transformations yield for the velocity

$$\left[\frac{d\mathbf{v}_I}{dt}\right]_I = \left[\frac{d\mathbf{v}_R}{dt}\right]_R + 2\Omega \times \mathbf{v}_R + \Omega \times (\Omega \times \mathbf{x}) + \frac{d\Omega}{dt} \times \mathbf{x},$$

where $\mathbf{x}$ is the position vector, and the three additional terms correspond to the *Coriolis acceleration*, the centripetal acceleration and the acceleration due to changes in the angular velocity itself, resp. By assuming a constant angular velocity and transformations to treat the centripetal acceleration as a force potential, included in the interior force term, the only remaining acceleration is the Coriolis term. With this, the momentum equation (7.7) in a rotating coordinate frame can be written

$$\frac{d\mathbf{v}_R}{dt} + 2\Omega \times \mathbf{v}_R = -\frac{1}{\rho}\nabla p + \nabla \Phi + \mathcal{F},$$

where Φ is a force potential, and $\mathcal{F}$ represents the other forces as above. One can show, that all terms on the right hand side (under certain assumptions, like Newtonian fluid, etc.) are independent of the frame of reference, and therefore, remain unchanged compared to (7.7).

The material derivatives of scalar quantities remain unchanged in a rotating frame of reference. However, the individual components of the material derivative are not independent. It holds

$$\left[\frac{d\eta}{dt}\right]_I = \left[\frac{d\eta}{dt}\right]_R - (\Omega \times \mathbf{x}) \cdot \nabla\eta,$$

and

$$\mathbf{v}_I \cdot \nabla\eta = (\mathbf{v}_R + \Omega \times \mathbf{x}) \cdot \nabla\eta,$$

for a scalar η. Following the presentation in [194], we start deriving the shallow water equations for rotating spherical geometries from the following set of equations, the momentum equation equipped with the Coriolis acceleration (as above) and the continuity equation from sect. 7.1:

$$\frac{d\mathbf{v}_R}{dt} = -\frac{1}{\rho}\nabla p - 2\Omega \times \mathbf{v}_R - \mathbf{g}, \tag{C.1}$$

$$\frac{dH}{dt} = -H\nabla \cdot \mathbf{v}_R. \tag{C.2}$$

We have simplified the force terms above to the internal gravitational force and used the notation $\mathbf{g}$ for the vector perpendicular to the earth's surface (in each point) of length g, the gravitational constant. With $\mathbf{e}_r$ the outward normal unit vector on the sphere, we can write $-\mathbf{g} = \mathbf{e}_r g$.

We use the following results for vector calculus on spherical manifolds [135]. Let $\xi = (\xi_1, \xi_2, \xi_3)$ be a vector in $\mathbb{R}^3$, then ξ_s denotes the tangential components of ξ at each point $\mathbf{x} \in S \subset \mathbb{R}^3$, where S is the 2D sphere in 3D space. We have

$$\xi_s = \xi - \mathbf{e}_r(\mathbf{e}_r \cdot \xi).$$

Furthermore, for the *tangential gradient* and the *tangential divergence* we have

$$\nabla_s \eta = \nabla\eta - \mathbf{e}_r(\mathbf{e}_r \cdot \nabla)\eta,$$
$$\nabla_s \cdot \xi = \nabla \cdot \xi - \mathbf{e}_r \cdot (\mathbf{e}_r \cdot \nabla)\xi.$$

The radial differential for a scalar η is defined by $\frac{\partial\eta}{\partial r} = \mathbf{e}_r \cdot \nabla\eta$. Using these notations we can immediately transform the material derivative for the tangential components of vectors on the sphere to obtain

$$\frac{d\xi_s}{dt} = \frac{\partial\xi_s}{\partial t} + (\mathbf{v} \cdot \nabla)\xi_s.$$

We define the *geopotential height* Φ by

$$\Phi = g(h - h_b) = gH.$$

With these tools at hand, we obtain the spherical form of the continuity equation (C.2):

$$\frac{d\Phi}{dt} + \Phi\nabla_s \cdot \mathbf{v}_s = 0. \tag{C.3}$$

In order to derive the tangential components of the spherical momentum equation we have to transform the Coriolis acceleration first. Additionally we use the Coriolis parameter $f_C = 2|\Omega|\sin\phi$, where ϕ is the latitude of the fluid element's position, in order to reflect the position dependent Coriolis acceleration on the surface of the sphere. After some additional scaling arguments and transformations that are not repeated here, we obtain the momentum equation of the spherical shallow water equations:

$$\frac{d\mathbf{v}_s}{dt} + \nabla_s\Phi + f_C(\mathbf{e}_r \times \mathbf{v}_s) = 0. \tag{C.4}$$

There are a number of different formulations of the spherical shallow water equations. We will report on some for reference and convenience. Williamson and coworkers [415] give several of the shallow water equations which we do not want to repeat here, since that reference is well known. Their formulations include the flux form SWE, the advective form of the SWE and the vorticity divergence form of the SWE, a stream function based formulation and a constrained 3D form. We want to mention some of the more recent formulations here. Note, that all of the mentioned formulations are pairwise equivalent. So, the choice of formulation depends very much on the numerical discretization scheme used.

We start with repeating the above result as a first form of the shallow water equations (SWE). In the following examples we will always put the momentum equation first.

Definition C.0.12 *(SWE with tangential operators and geopotential height)*

$$\frac{d\mathbf{v}_s}{dt} + \nabla_s\Phi + f_C(\mathbf{e}_r \times \mathbf{v}_s) = 0$$

$$\frac{d\Phi}{dt} + \Phi\nabla_s \cdot \mathbf{v}_s = 0$$

Heinze and Hense combine this form of the SWE with a splitting of Φ into a constant term and a fluctuation [194]. The second form follows the presentation in [273], and uses the fluid layer height instead of the geopotential height.

Definition C.0.13 *(SWE with fluid layer height)*

$$\frac{d\mathbf{v}_s}{dt} + g\nabla h + f_C(\mathbf{e}_r \times \mathbf{v}_s) = 0,$$

$$\frac{dH}{dt} + H\nabla \cdot \mathbf{v}_s = 0.$$

The conservative form of the shallow water equations as in [273]:

Definition C.0.14 *(SWE in conserving form according to Majda)*

$$\frac{\partial (H\mathbf{v}_s)}{\partial t} + \nabla \cdot \left(H\mathbf{v}_s \otimes \mathbf{v}_s + \frac{g}{2}H^2 I \right) = -(f_C \mathbf{e}_r \times \mathbf{v}_s + g\nabla h_B)H,$$

$$\frac{\partial H}{\partial t} + \nabla \cdot (H\mathbf{v}_s) = 0.$$

We have used $\otimes$ as the tensor product. If we assume a flat bottom, a plane domain, and no rotation, the right hand side of the momentum equation in definition C.0.14 becomes zero and the shallow water equations are also called *isentropic fluid flow equations*.

Giraldo uses a 3D cartesian coordinate formulation of the shallow water equations and introduces Lagrange multipliers to restrict the velocity vectors to the sphere's surface in [167]. We cite his compact form of the shallow water equations:

Definition C.0.15 *(SWE in Cartesian coordinates according to Giraldo)*
The SWE are given as a system of equations, defined by

$$\frac{\partial \mathbf{\Phi}}{\partial t} + \nabla \cdot (\mathbf{\Phi}\mathbf{v}^T) = \mathbf{S}(\mathbf{\Phi}),$$

with

$$\mathbf{\Phi} = \begin{bmatrix} \Phi \\ \Phi u \\ \Phi v \\ \Phi w \end{bmatrix}, \quad \mathbf{v} = \begin{bmatrix} u \\ v \\ w \end{bmatrix}, \quad \mathbf{S}(\mathbf{\Phi}) = -\Phi\nabla\Phi - f_C(\mathbf{e}_r \times \Phi\mathbf{u}) + \mu\mathbf{e}_r.$$

The last term in $\mathbf{S}(\mathbf{\Phi})$ *is a force that restricts the 3D velocity vector to the sphere's surface with a Lagrange multiplier* μ *determined in the discretized equations.*

Ringler and Randall use a formulation that includes the kinetic energy and absolute vorticity [339]:

Definition C.0.16 *(SWE with kinetic energy and absolute vorticity)*

$$\frac{\partial \mathbf{v}}{\partial t} + \zeta\mathbf{e}_r \times \mathbf{v} + \nabla[k + g(h + h_s)] = 0,$$

$$\frac{\partial h}{\partial t} + \nabla \cdot (h\mathbf{v}) = 0,$$

with h_s *the fluid surface height,* h *the fluid layer thickness,* k *the kinetic energy, and* ζ *the absolute vorticity,*

$$\zeta = f_C + \mathbf{e}_r \cdot \nabla \times \mathbf{v}.$$

A scalar formulation is also given in [339]:

Definition C.0.17 *(Scalar SWE with kinetic energy and absolute vorticity)*

$$\frac{\partial \delta}{\partial t} - \mathbf{e}_r \cdot \nabla \times (\zeta \mathbf{v}) + \Delta[k + g(h + h_s)] = 0,$$

$$\frac{\partial \zeta}{\partial t} + \nabla \cdot (\zeta \mathbf{v}) = 0,$$

$$\frac{\partial h}{\partial t} + \nabla \cdot (h \mathbf{v}) = 0,$$

with $\delta = \nabla \cdot \mathbf{v}$ the divergence, and ζ the absolute vorticity. The velocity vector has to be reconstructed by a Helmholtz decomposition.

A scalar form using the vorticity-divergence formulation with a Helmholtz decomposition is used by Läuter [255] (see sect. 8.3).

Definition C.0.18 *(SWE in vorticity divergence form with tangential operators)*

$$\frac{d\zeta}{dt} + \zeta \delta + \delta f_C + \mathbf{v} \cdot \nabla_s f_C = 0,$$

$$\frac{d\delta}{dt} + \Delta_s \Phi - \zeta f_C + (\mathbf{e}_r \times \mathbf{v}) \cdot \nabla_s f_C + J(\mathbf{v}) = 0,$$

$$\frac{d\Phi}{dt} + \delta(\Phi - \Phi_0) - \mathbf{v} \cdot \nabla_s \Phi_0 = 0.$$

Note that the operators ∇_s, and Δ_s together with the material derivative $\frac{d}{dt}$ are tangential operators. J is a matrix specified in sect. 8.3, $\zeta = \nabla_s \times \mathbf{v}$ is the vorticity and $\delta = \nabla_s \cdot \mathbf{v}$ the divergence, Φ_0 the orography.

Yet another scalar formulation with the fluid height, the potential vorticity and the divergence is used by Pesch [316].

Definition C.0.19 *(SWE in height divergence form with tangential operators)*

$$\frac{\partial \delta}{\partial t} + \nabla \cdot \left[h\zeta \mathbf{e}_r \times \mathbf{v} + \nabla \left(g(h + h_s) + \frac{1}{2} \mathbf{v} \cdot \mathbf{v} \right) \right] = 0,$$

$$\frac{\partial h}{\partial t} + \nabla \cdot (h \mathbf{v}) = 0,$$

$$\frac{\partial h\zeta}{\partial t} + \nabla \cdot (h \zeta \mathbf{v}) = 0.$$

Pesch uses the potential vorticity ζ.

Blikberg uses the shallow water equations in spherical geometry in the following compact form [63]:

Definition C.0.20 *(SWE in spherical compact form)*

$$\frac{\partial \mathbf{q}}{\partial t} + \partial_\phi \mathbf{f} + \partial_\lambda \mathbf{g} = \mathbf{s},$$

where (λ, ϕ) are the latitudinal and longitudinal coordinate pair, ∂_ν is the partial derivative in ν direction and

$$
\mathbf{q} = \begin{bmatrix} \hat{h} \\ \hat{h}u \\ \hat{h}v \end{bmatrix}, \quad \mathbf{f} = \begin{bmatrix} a(\hat{h}u) \\ a(\hat{h}u^2) + b(\hat{h})^2 g \\ a(\hat{h}uv) \end{bmatrix}, \quad \mathbf{g} = \begin{bmatrix} c(\hat{h}v) \\ c(\hat{h}uv) \\ c(\hat{h}v^2) + d(\hat{h})^2 g \end{bmatrix},
$$

with $\hat{h} = h \cos \lambda$, $a = (r_{\text{earth}} \cos \lambda)^{-1}$, $b = (2r_{\text{earth}} \cos^2 \lambda)^{-1}$, $c = r_{\text{earth}}^{-1}$, and $d = (2r_{\text{earth}} \cos \lambda)^{-1}$. The right hand side is given by

$$
\mathbf{s} = \begin{bmatrix} 0 \\ (f_C + cu \tan \lambda)\hat{h}v \\ -(f_C + cv \tan \lambda)\hat{h}u - d(\hat{h})^2 g \tan \lambda \end{bmatrix}.
$$

Note that this set of equations does not involve orography.

D

List of Notations

Notation	Description	First occurrence
$I = [0, T]$	time interval	2.3
$\mathcal{G} \in \mathbb{R}^d$, $\mathcal{G}_h \in \mathbb{R}^d$	continuous and discrete computational domain	2.3
ρ_h	discrete (scalar) variable corresponding to ρ	2.3
τ	element of a triangulation	2.3
h	mesh size parameter, also used as discrete subscript	2.4
η, η_τ	(local) refinement criterion	2.3
θ_{ref}, θ_{crs}	refinement/coarsening tolerance	2.4
$\mathcal{T}$	a triangulation	2.4
$\mathcal{T}_1 \prec \mathcal{T}_2$	hierarchically refined triangulations	3.1.7
ϑ	inner angle of triangulation	3.1
v_i	vertex of a triangulation	3.3
τ^{3D}	element/tetrahedron in 3D triangulation	3.4
(λ, ϕ, r)	spherical coordinates	3.5
∂_x, ∂_t	first order partial differential operators	4.2.1
Δ	Lalpace's operator	4.2.2
$\mathcal{L}$, $\mathcal{E}$, $\mathcal{R}$	Load balancing, edge cut, and re-partitioning cost	B
$G(V, E)$	graph with vertex set V and edges E	5.1
Δx, Δt	spatial and time step	3.5, 4.2.1
$\cdot\vert_x$, $\cdot\vert_y$	x- and y-component of $(\cdot)$	6.1.1
b, b_i	usually a basis function	6.1.3
$\mathcal{V}$, $\mathcal{V}_h$	continuous and discrete function space	6.1.3
ε, $[\varepsilon]$	true error and error estimate	2.4, 2.6.3
δ_{ij}	Dirac delta function	6.1.3
$\mathcal{D}$, $\mathcal{D}^*$	abstract differential operator, and adjoint	6.1.3, 8.2
φ	radial basis function	6.1.4

Notation continued:

Notation	Description	First occurence
$\check{\mathbf{x}}$	trajectory or particle path	A
$V(t)$, V	material volume, control volume	A, 7.2.2
$\frac{d}{dt}$	material derivative	A
$\mathbf{n}$	outward unit normal	A.0.1
σ_{surf}	surface stress tensor	7.1.2
$\mathcal{F}$	force term	7.1.2
Ω	angular velocity	C
$\mathbf{v}$	velocity (vector)	C
Φ	geopotential height	C
f_C	Coriolis parameter	C
∇_s, Δ_s	tangential gradient/divergence, and Laplacian	C, 8.3.1

References

1. R. Abgrall and A. Harten, *Multiresolution representation in unstructured meshes*, SIAM J. Numer. Anal. **35** (1998), no. 6, 2128–2146.
2. R. A. Adams and J. F. J. Fournier, *Sobolev spaces*, 2nd ed., Academic Press, Amsterdam, 2003.
3. G. Adrian, *Parallel processing in regional climatology: The parallel version of the "Karlsruhe Atmospheric Mesoscale Model" (KAMM)*, Parallel Computing **25** (1999), 777–787.
4. M. Aftosmis, M. Berger, J. Melton, and S. Murman, *Cart3D homepage*, http://people.nas.nasa.gov/%7Eaftosmis/cart3d/.
5. H. Akima, *Algorithm 526: Bivariate interpolation and smooth surface fitting for irregularly distributed data points*, ACM Trans. on Math. Softw. **4** (1978), no. 2, 160–164.
6. A. S. Almgren, J. B. Bell, P. Collela, L. H. Howell, and M. L. Welcome, *A conservative adaptive projection method for the variable density incompressible navier-stokes equations*, Journal of Computational Physics **142** (1998), 1–46.
7. T. Alsos, *Effective ODE-solvers for trajectory calculations in semi-Lagrangian methods used in weather forcasting*, Diploma thesis, Department of Mathematical Sciences, The Norwegian University of Science and Technology, Trondheim, Norway, 1998.
8. P. R. Amestoy, T. A. Davis, and I. S. Duff, *Algorithm 837: AMD, an approximate minimum degree ordering algorithm*, ACM Trans. on Math. Software **30** (2004), no. 3, 381–388.
9. I. O. Angell, *High resolution computer graphics using C*, Macmillan Computer Science Series, Macmillan, Basingstoke, Hampshire, 1990.
10. T. Arbogast and M. F. Wheeler, *A characteristics-mixed finite element method for advection-dominated transport problems*, SIAM J. Numer. Anal. **32** (1995), no. 2, 404–424.
11. P. Arminjon and A. St-Cyr, *Nessyahu-Tadmor-type central finite volume methods without predictor for 3d Cartesian and unstructured tetrahedral grids*, App. Numer. Math. **46** (2003), 135–155.
12. D. N. Arnold, F. Brezzi, B. Cockburn, and L. D. Marini, *Unified analysis of discontinuous Galerkin methods for elliptic problems*, SIAM J. Numer. Anal. **39** (2002), no. 5, 1749–1779.

13. B. N. Azarenok and T. Tang, *Second-order Godunov-type scheme for reactive flow calculations on moving meshes*, J. Comput. Phys. **206** (2005), 48–80.

14. Després B., *An explicit a priori estimate for a finite volume approximation of linear advection on non-Cartesian grids*, SIAM J. Numer. Anal. **42** (2004), no. 2, 484–504.

15. I. Babuska, *The selfadaptive approach in the finite element method*, The Mathematics of Finite Elements and Applications, 1976, Proceedings of the Brunel University Conference of the Institute of Mathematics and its Applications held in April 1975 (MAFELAP 1975), pp. 125–142.

16. I. Babuska and M. R. Dorr, *Error estimates for the combined h and p versions of the finite element method*, Numer. Math. **37** (1981), 257–277.

17. I. Babuska and W. C. Rheinboldt, *A posteriori error estimates for the finite element method*, Int. J. Numer. Meth. Eng. **12** (1978), 1597–1615.

18. I. Babuška and W. C. Rheinboldt, *Error estimates for adaptive finite element computations*, SIAM J. Numer. Anal. **15** (1978), no. 4, 736–754.

19. I. Babuska, B. A. Szabo, and I. N. Katz, *The p-verions of the finite element method*, SIAM J. Numer. Anal. **18** (1981), no. 3, 515–545.

20. D. P. Bacon, N. N. Ahmad, Z. Boybeyi, T. J. Dunn, M. S. Hall, P. C. S. Lee, R. A. Sarma, M. D. Turner, K. T. Wraight III, S. H. Young, and J. W. Zack, *A dynamically adapting weather and dispersion model: The operational multiscale environment model with grid adaptivity (OMEGA)*, Mon. Wea. Rev. **128** (2000), 2044–2076.

21. R. E. Bank and Holst M., *A new paradigm for parallel adaptive meshing algorithms*, SIAM J. Sci. Comput. **22** (2000), no. 4, 1411–1443.

22. R. E. Bank and A. Weiser, *Some a posteriori error estimators for elliptic partial differential equations*, Math. Comp. **44, No. 170** (1985), 283–301.

23. R. E. Bank and J. Xu, *Asymptotically exact a posteriori error estimators, part I: Grids with superconvergence*, SIAM J. Numer. Anal. **41** (2003), no. 6, 2294–2312.

24. _____, *Asymptotically exact a posteriory error estimators, part II: General unstructured grids*, SIAM J. Numer. Anal. **41** (2003), no. 6, 2313–2332.

25. E. Bänsch, *Local mesh refinement in 2 and 3 dimensions*, Impact of Comput. in Sci. and Eng. **3** (1991), 181–191.

26. S. R. M. Barros and C. I. Garcis, *A global semi-implicit semi-Lagrangian shallow-water model on locally refined grids*, Mon. Wea. Rev. **132** (2004), 53–65.

27. P. Bartello, *A comparison of time discretization schemes for two-timescale problems in geophysical fluid dynamics*, J. Comput. Phys. **179** (2002), 268–285.

28. T. J. Barth, *Simplified discontinuous Galerkin methods for systems of conservation laws with convex extensions*, Discontinuous Galerkin Methods (Berlin, Heidelberg, New York) (B. Cockburn, G. E. Karniadakis, and C.-W. Shu, eds.), Lecture Notes in Computational Science and Engineering, vol. 11, Springer Verlag, 2000, pp. 63–75.

29. J. Baudisch, *Accurate reconstruction of vector fields using radial basis functions*, Thesis, Technische Universität München, Boltzmannstr. 3, 85747 Garching, Germany, 2005.

30. J. R. Baumgardner and P. O. Frederickson, *Icosahedral discretization of the two-sphere*, SIAM J. Numer. Anal. **22** (1985), no. 6, 1107–1115.

31. M. Bause and P. Knabner, *Uniform error analysis for Lagrange-Galerkin approximations of convection-dominated problems*, SIAM J. Numer. Anal. **39** (2002), no. 6, 1954–1984.

32. J. Behrens, *amatos – Adaptive mesh generator for atmospheric and oceanic simulation*,
    http://www.amatos.info.

33. _____, *Adaptive Semi-Lagrange-Finite-Elemente-Methode zur Lösung der Flachwassergleichungen: Implementierung und Parallelisierung*, Ber. Polarforsch. **217** (1996).

34. _____, *An adaptive semi-Lagrangian advection scheme and its parallelization*, Mon. Wea. Rev. **124** (1996), no. 10, 2386–2395.

35. _____, *A parallel adaptive finite-element semi-Lagrangian advection scheme for the shallow water equations*, Modeling and Computation in Environmental Sciences (Braunschweig) (R. Helmig, W. Jäger, W. Kinzelbach, P. Knabner, and G. Wittum, eds.), Notes on Numerical Fluid Mechanics, vol. 59, Vieweg, 1997, Proceedings of the First GAMM-Seminar at ICA Stuttgart, October 12–13, 1995, pp. 49–60.

36. _____, *Atmospheric and ocean modeling with an adaptive finite element solver for the shallow-water equations*, Applied Numerical Mathematics **26** (1998), no. 1–2, 217–226.

37. J. Behrens, *amatos – Adaptive mesh generator for atmosphere and ocean simulation*, Technische Universität München, TUM, Center for Mathematical Sciences, D-80290 Munich, Germany, 2002, API Documentation Version 1.2.

38. J. Behrens, *Adaptive mesh generator for atmospheric and oceanic simulations – amatos*, Technical Report TUM-M0409, Technische Universität München, Zentrum Mathematik, Boltzmannstr. 3, 85747 Garching, 2004,
    http://www-lit.ma.tum.de/veroeff/html/040.65008.html.

39. J. Behrens, K. Dethloff, W. Hiller, and A. Rinke, *Evolution of small-scale filaments in an adaptive advection model for idealized tracer transport*, Mon. Wea. Rev. **128** (2000), 2976–2982.

40. J. Behrens and A. Iske, *Grid-free adaptive semi-Lagrangian advection using radial basis functions*, Comp. Math. Appl. **43** (2002), 319–327.

41. J. Behrens, A. Iske, and S. Pöhn, *Effective node adaption for grid-free semi-Lagrangian advection*, Discrete Modelling and Discrete Algorithms in Continuum Mechanics (Berlin) (Th. Sonar and I. Thomas, eds.), Logos Verlag, 2001, Proceedings of the GAMM Workshop, November 24-25, 2000, Technical University of Brunswick, Germany, pp. 110–119.

42. J. Behrens and L. Mentrup, *A conservative scheme for 2D and 3D adaptive semi-Lagrangian advection*, Recent Advances in Adaptive Computation (Providence, Rhode Island) (Z.-C. Shi, Z. Chen, T. Tang, and D. Yu, eds.), Contemporary Mathematics, vol. 383, American Mathematical Society, 2005, pp. 219–233.

43. J. Behrens, N. Rakowsky, W. Hiller, D. Handorf, M. Läuter, J. Päpke, and K. Dethloff, *amatos: Parallel adaptive mesh generator for atmospheric and oceanic simulation*, Technical Report TR 02-03, BremHLR – Competence Center of High Performance Computing Bremen, Bremen, Germany, 2003,
    http://www.bremhlr.uni-bremen.de/TR_0203.pdf.

44. _____, *amatos: Parallel adaptive mesh generator for atmospheric and oceanic simulation*, Ocean Modelling **10** (2005), no. 1–2, 171–183.

45. J. Behrens and J. Zimmermann, *Parallelizing an unstructured grid generator with a space-filling curve approach*, Euro-Par 2000 Parallel Processing – 6th International Euro-Par Conference Munich, Germany, August/Sptember 2000 Proceedings (Berlin) (A. Bode, T. Ludwig, W. Karl, and R. Wismüller, eds.), Lecture Notes in Computer Science, vol. 1900, Springer Verlag, 2000, pp. 815–823.

46. M. J. Bell, *Conservation of potential vorticity on Lorenz grids*, Mon. Wea. Rev. **131** (2003), 1498–1501.

47. C. Belwal, A. Sandu, and E. M. Constantinescu, *Adaptive resolution modeling of regional air quality*, Proceedings of the 2004 ACM symposium on Applied computing (New York), ACM Symposium on Applied Computing, ACM Press, 2004,
http://portal.acm.org/citation.cfm?id=967900.967951, pp. 235–239.

48. P. Bénard, *Stability of semi-implicit and iterative centered-implicit time discretizations for various equations systems used in NWP*, Mon. Wea. Rev. **131** (2003), 2479–2491.

49. _____, *On the use of a wider class of linear systems for the design of constant-coefficients semi-implicit time schemes in NWP*, Mon. Wea. Rev. **132** (2004), 1319–1324.

50. P. Bénard, R. Laprise, J. Vivoda, and P. Smolíková, *Stability of leapfrog constant-coefficients semi-implicit schemes for the fully elastic system of Euler equations: Flat-terrain case*, Mon. Wea. Rev. **132** (2004), 1306–1318.

51. M. J. Berger, M. J. Aftosmis, D. D. Marshall, and S. M. Murman, *Performance of a new CFD solver using a hyprid programming paradigm*, J. Parallel Distrib. Comput. **65** (2005), 414–423.

52. M. J. Berger and P. Colella, *Local adaptive mesh refinement for shock hydrodynamics*, Jou. Comp. Phys. **82** (1989), 64–84.

53. M. J. Berger, C. Helzel, and R. J. LeVeque, *h-box methods for the approximation of hyperbolic conservation laws on irregular grids*, SIAM J. Numer. Anal. **41** (2003), no. 3, 893–918.

54. M. J. Berger and R. J. LeVeque, *Adaptive mesh refinement using wave-propagation algorithms for hyperbolic systems*, SIAM J. Numer. Anal. **35** (1998), no. 6, 2298–2316.

55. M. J. Berger and J. Oliger, *Adaptive mesh refinement for hyperbolic partial differential equations*, Jou. Comp. Phys. **53** (1984), 484–512.

56. R. Bermejo, *A Galerkin-characteristic algorithm for transport-diffusion equations*, SIAM J. Numer. Anal. **32** (1995), no. 2, 425–454.

57. R. Bermejo and J. Conde, *A conservative quasi-monotone semi-Lagrangian scheme*, Mon. Wea. Rev. **130** (2002), 423–430.

58. R. Bermejo and A. Staniforth, *The conversion of semi-Lagrangian advection schemes to quasi-monotone schemes*, Mon. Wea. Rev. **120** (1992), 2622–2632.

59. G. Berti, *A calculus for stencils on arbitrary grids with applications to parallel PDE solution*, Proceedings of the GAMM Workshop Discrete Modelling and Discrete Algorithms in Contin (Braunschweig) (T. Sonar and I. Thomas, eds.), Logos Verlag, 2001.

60. N. Biggs, *Algebraic graph theory*, 2 ed., Cambridge University Press, Cambridge, 1994.

61. E. Blayo and L. Debreu, *Adaptive mesh refinement for finite difference ocean models: Some first experiments*, Project IDOPT, Laboratoire de Modélisation et Calcul, Université Joseph Fourier, Grenoble, France, 1998.

62. E. Blayo, L. Debreu, G. Mounié, and D. Trystram, *Dynamic load balancing for ocean circulation model with adamtive meshing*, Euro-Par 1999 (Berlin, Heidelberg, New York) (P. Amestoy, P. Berger, M. Daydé, I. Duff, V. Fraysse, L. Giraud, and D. Ruiz, eds.), Lecture Notes in Computer Science, vol. 1685, Springer Verlag, 1999, pp. 303–312.

63. R. Blikberg, *Nested parallelism in OpenMP with application to adaptive mesh refinement*, Phd thesis, Parallab/Department of Informatics, University of Bergen, Bergen, Norway, 2003.

64. G. J. Boer and B. Denis, *Numerical convergence of the dynamics of a GCM*, Climate Dynamics **13** (1997), 359–374.

65. L. Bonaventura and G. Rosatti, *A cascadic conjugate gradient algorithm for mass conservative, semi-implicit discretization of the shallow water equations on locally refined structured grids*, Int. J. Numer. Meth. Fluids **40** (2002), 217–230.

66. F. A. Bornemann and P. Deuflhard, *The cascadic multigrid method for elliptic problems*, Numer. Math. **75** (1996), 135–152.

67. N. Botta, R. Klein, S. Langenberg, and S. Lützenkirchen, *Well balanced finite volume methods for nearly hydrostatic flows*, J. Computat. Phys. **196** (2004), 539–565.

68. A. Bourchtein, *Semi-Lagrangian semi-implicit space splitting regional baroclinic atmospheric model*, App. Num. Math. **40** (2002), 307–326.

69. Z. Boybeyi, N. N. Ahmad, D. P. Bacon, T. J. Dunn, M. S. Hall, P. C. S. Lee, R. A. Sarma, and T. R. Wait, *Evaluation of the operational multiscale environment model with grid adaptivity against the European Tracer Experiment*, Journal of Applied Meteorology **40** (2001), no. 9, 1541–1558.

70. J. U. Brackbill and J. S. Saltzman, *Adaptive zoning for singular problems in two dimensions*, J. Comput. Phys. **46** (1982), 342–368.

71. D. Braess, *Finite elements – theory, fast solvers, and applications in solid mechanics*, 2nd ed., Cambridge University Press, Cambridge, UK, 2001.

72. A. Bregman, A. F. J. van Velthoven, F. G. Wienhold, H. Fischer, T. Zenker, A. Waibel, A. Frenzel, F. Arnold, G. W. Harris, M. J. A. Bolder, and J. Lelieveld, *Aircraft measurements of O_3, HNO_3, and N_2O in the winter arctic lower stratosphere during the stratosphere-troposphere experiment by aircraft measurements (STREAM) 1*, J. Geophys. Res. **100** (1995), 11,245–11,260.

73. G. Breinholt and C. Schierz, *Algorithm 781: Generating Hilbert's space-filling curve by recursion*, AMS Trans. Math. Softw. **24** (1998), 184–189.

74. G. L. Browning, J. J. Hack, and P. N. Swarztrauber, *A comparison of three numerical methods for solving differential equations on the sphere.*, Mon. Wea. Rev. **117** (1989), 1058–1075.

75. M. D. Buhmann, *Radial basis functions: Theory and implementations*, Cambridge University Press, 2003.

76. A. C. Calder, B. C. Curtis, L. J. Dursi, B. Fryxell, G. Henry, P. MacNeice, K. Olson, P. Ricker, R. Rosner, F. X. Timmes, H. M. Tufo, J. W. Truran, and M. Zingale, *High-performance reactive fluid flow simulations using adaptive mesh refinement on thousands of processors*, Proceedings of the 2000 ACM/IEEE conference on Supercomputing (CDROM), IEEE Computer Society, 2000,
http://csdl.computer.org/comp/proceedings/sc/2000/9802/00/98020056abs.htm.

77. W. Cao, W. Huang, and R. D. Russell, *Approaches for generating moving adaptive meshes: Locality versus velocity*, App. Numer. Math. **47** (2003), 121–138.

78. C. Carstensen, *All first-order averaging techniques for a posteriori finite element error control on unstructured grids are efficient and reliable*, Math. Comp. **73** (2004), no. 247, 1153–1165.

79. C. Caruso and F. Quarta, *Interpolation methods comparison*, Computers Math. Applic. **35** (1998), no. 12, 109–126.

80. M. A. Celia, T. F. Russell, I. Herrera, and R. E. Ewing, *An Eulerian-Lagrangian localized adjoint method for the advection-diffusion equation*, Adv. Water Resources **13** (1990), 187–206.

81. S. Chen, E. Weinan, and C.-W. Shu, *The heterogeneous multi-scale method based on the discontinuous Galerkin method for hyperbolic and parabolic problems*, http://www.dam.brown.edu/scicomp/publications/Reports/Y2004/BrownSC-2004-11.pdf, 2004.

82. S.-H. Chen and W.-Y. Sun, *Application of the multigrid method and a flexible hybrid coordinate in a nonhydrostatic model*, Mon. Wea. Rev. **129** (2001), 2660–2676.

83. Z. Chen, *Characteristic-nonconforming finite-element methods for advection-dominated diffusion problems*, Comp. Math. Appl. **48** (2004), 1087–1100.

84. G. Chesshire and W. D. Henshaw, *Composite overlapping meshes for the solution of partial differential equations*, J. Comput. Phys. **90** (1990), 1–64.

85. B.-J. Choi, M. Iskandarani, J. Levin, and D. B. Haidvogel, *A spectral finite-volume method for the shallow water equations*, Mon. Wea. Rev. **132** (2004), 1777–1791.

86. H.-W. Choi and M. Paraschivoiu, *A posteriori finite element output bounds with adaptive mesh refinement: Application to a heat transfer problem in a three-dimensional rectangular duct*, Comput. Methods Appl. Mech. Engrg. **191** (2002), 4905–4925.

87. _____, *Adaptive computations of a posteriori finite element output bounds: A comparison of the "hybrid-flux" approach and the "flux-free" approach*, Comput. Methods Appl. Mech. Engrg. **193** (2004), 4001–4033.

88. N. Chrisochoides, *Parallel grid generation*, http://www.npac.syr.edu/PROJECTS/PUB/nikos/ParGG.html, 1994.

89. P. G. Ciarlet, *The finite element method for elliptic problems*, North–Holland Publishing Co., Amsterdam, 1978.

90. R. W. Clough, *Original formulation of the finite element method*, Finite Elements in Analysis and Design **7** (1990), 89–101.

91. R. Cocci Grifoni, F. Bisegna, and G. Passerini, *A refinement of AERMOD results by means of mesoscale model simulation*, Proceedings from the 8th International Conference on Harmonisation within Atmospheric Dispersion Modelling for Regulatory Purposes (Sofia), October 2002, http://www.harmo.org/conferences/proceedings/_Sofia/Sofia_proceedings.asp, pp. 191–195.

92. B. Cockburn, *Devising discontinuous Galerkin methods for non-linear hyperbolic conservation laws*, J. Comput. Appl. Math. **128** (2001), 187–204.

93. _____, *Discontinuous Galerkin methods*, Z. Angew. Math. Mech. **83** (2003), no. 11, 731–754.

94. B. Cockburn, F. Coquel, and P. G. LeFloch, *Convergence of the finite volume method for multidimensional conservation laws*, SIAM J. Numer. Anal. **32** (1995), no. 3, 687–705.

95. B. Cockburn, G. E. Karniadakis, and C.-W. Shu, *The development of discontinuous Galerkin methods*, Discontinuous Galerkin Methods (Berlin, Heidelberg, New York) (B. Cockburn, G. E. Karniadakis, and C.-W. Shu, eds.), Lecture Notes in Computational Science and Engineering, vol. 11, Springer Verlag, 2000, pp. 3–50.

96. B. Cockburn and C.-W. Shu, *The local discontinuous Galerkin method for time-dependent convection-diffusion systems*, SIAM J. Numer. Anal. **35** (1998), no. 6, 2440–2463.

97. P. Colella, D. T. Graves, T. J. Ligocki, D. F. Martin, D. Modiano, D. B. Serafini, and B. Van Straalen, *CHOMBO homepage*, http://seesar.lbl.gov/anag/chombo.

98. _____, *Chombo software package for AMR applications – design document*, Lawrence Berkeley National Laboratory, Applied Numerical Algorithms Group, NERSC Division, Berkeley, CA, 2003.

99. G. Corliss, C. Faure, A. Griewank, L. Hascoet, and U. Naumann (eds.), *Automatic differentiation of algorithms: From simulation to optimization*, Springer Verlag, New York, 2002.

100. J. Côté, *A Lagrange multiplier approach for the metric terms of semi-Lagrangian models on the sphere*, Quart. J. Roy. Meteor. Soc. **114** (1988), 1347–1352.

101. J. Côté, S. Gravel, A. Méthot, A. Patoine, M. Roch, and A. Staniforth, *The operational CMC-MRB global environmental multiscale (GEM) model: Part I – design considerations and formulation*, Mon. Wea. Rev. **126** (1998), no. 6, 1373–1395.

102. R. Courant, K. O. Friedrichs, and H. Lewy, *Über die partiellen Differenzengleichungen der mathematischen Physik*, Math. Annalen **100** (1928), 67–108.

103. E. N. Curchitser, M. Iskandarani, and D. B. Haidvogel, *A spectral element solution of the shallow water equations on multiprocessor computers*, Preprint, 1996.

104. E. Cuthill and J. McKee, *Reducing the bandwidth of sparse symmetric matrices*, ACM/CSC-ER Proceedings of the 1969 24th national conference, ACM Press New York, NY, USA, 1969, pp. 157–172.

105. Sridar. D. and N. Balakrishnan, *An upwind finite difference scheme for meshless solvers*, J. Comput. Phys. **189** (2003), 1–29.

106. S. Danilov, *A brief description of finite-element shallow water model*, personal communication, 2005.

107. S. Danilov, G. Kivman, and J. Schröter, *A finite-element ocean model: Principles and evaluation*, Ocean Modelling **6** (2004), 125–150.

108. J. R. Davis and Y. P. Sheng, *Development of a parallel storm surge model*, Int. J. Numer. Meth. Fluids **42** (2003), 549–580.

109. T. A. Davis, J. R. Gilbert, S. G. Larimore, and E. G. Ng, *Algorithm 836: COLAMD, a column approximate minimum degree ordering algorithm*, ACM Trans. on Math. Software **30** (2004), no. 9, 377–380.

110. T. A. Davis, J. R. Gilbert, S. I. Larimore, and E. G. Ng, *A column approximate minimum degree ordering algorithm*, ACM Trans. on Math. Software **30** (2004), no. 3, 353–376.

111. M. de Berg, M. van Kreveld, M. Overmars, and O. Schwarzkopf, *Computational geometry: Algorithms and applications*, 2nd revised ed., Springer, Berlin, Heidelberg, New York, 2000.

112. A. Dedner and P. Vollmöller, *An adaptive higher order method for solving the radiation transfport equation on unstructured grids*, J. Comput. Phys. **178** (2002), 263–289.

113. R. Deiterding, *AMROC homepage*, http://amroc.sourceforge.net.

114. _____, *Parallel adaptive simulation of multi-dimensional detonation structures*, PhD thesis, Brandenburgische Technische Universität Cottbus, Cottbus, Germany, 2003.

115. E. D. Dendy, N. T. Padial-Collins, and W. B. VanderHeyden, *A general-purpose finite-volume advection scheme for continuous and discontinuous fields on unstructured grids*, J. Comput. Phys. **180** (2002), 559–583.

116. J. M. Dennis, *Partitioning with space-filling curves on the cubed-sphere*, http://www.scd.ucar.edu/css/publications/sfc3.pdf, 2003.

117. M. Déqué, C. Dreverton, A. Braun, and D. Cariolle, *The ARPEGE/IFS atmosphere model: a contribution to the French community climate modelling*, Climate Dynamics **10** (1994), no. 4–5, 249–266.

118. L. DeRose, K. Gallivan, E. Galloopoulos, and A. Navarra, *Parallel ocean circulation modeling on CEDAR*, CSRC Report 1124, University of Illinois at Urbana-Champaign, Center for Supercomputing Research and Development, Urbana, Illinois, 1991.

119. K. Dethloff, A. Rinke, R. Lehmann, J. H. Christensen, M. Botzet, and B. Machenhauer, *Regional climate model of the arctic atmosphere*, Jou. Geophys. Research **101** (1996), no. D18, 23401–23422.

120. P. Deuflhard, *Recent progress in extrapolation methods for ordinary differential equations*, SIAM Review **27** (1985), no. 4, 505–535.

121. P. Deuflhard and F. Bornemann, *Numerische Mathematik II*, de Gruyter, 2002.

122. P. Deuflhard, P. Leinen, and H. Yserentant, *Concepts of an adaptive hierarchical finite element code*, IMPACT Comp. Sci. Eng. **1** (1989), no. 1, 3–35.

123. R. Diekmann, A. Frommer, and B. Monien, *Efficient schemes for nearest neighbor load balancing*, Parallel Computing **25** (1999), 789–812.

124. R. Diekmann, D. Meyer, and B. Monien, *Parallel decomposition of unstructured FEM-meshes*, Proceedings of IRREGULAR 95, Lecture Notes in Computer Science, vol. 980, Springer-Verlag, 1995, pp. 199–215.

125. G. S. Dietachmayer and K. K. Droegemeier, *Application of continuous dynamic grid adaptation techniques to meteorological modeling. Part I: Basic fromulation and accuracy*, Mon. Wea. Rev. **120** (1992), 1675–1706.

126. U. Dobrind, *Ein Inversmodell für den Südatlantik mit der Methode der finiten Elemente*, PhD thesis, Universität Bremen, Bremen, Germany, 1999, http://elib.suub.uni-bremen.de/publications/dissertations/E-Diss35_Dobrindt_U1999.pdf.

127. J. J. Dongarra, I. S. Duff, D. C. Sorensen, and H. A. van der Vorst, *Numerical linear algebra for high-performance computers*, SIAM, Philadelphia, 1998.

128. W. Dörfler, *A convergent adaptive algorithm for Poisson's equation*, SIAM J. Numer. Anal. **33** (1996), no. 3, 1106–1134.

129. J. Douglas Jr. and T. F. Russell, *Numerical methods for convection-dominated diffusion problems based on combining the method of characteristics with finite*

element or finite difference procedures, SIAM J. Numer. Anal. **19** (1982), no. 5, 871–885.

130. D. G. Dritschel and M. H. P. Ambaum, *A contour-advective semi-Lagrangian numerical algorithm for simulating fine-scale conservative dynamical fields*, Q. J. R. Meteorol. Soc. **123** (1997), 1097–1130.

131. J. K. Dukowicz, *Mesh effects for Rossby waves*, J. Comput. Phys. **119** (1995), 188–194.

132. M. Dumbser and C.-D. Munz, *Arbitrary high order discontinuous Galerkin schemes*, http://www.iag.uni-stuttgart.de/people/michael.dumbser/files/PAPER_ADER-DG.pdf, 2003.

133. P. A. Durbin and G. Iaccarino, *An approach to local refinement of structured grids*, J. Comput. Phys. **181** (2002), 639–653.

134. J. A. Dutton, *Dynamics of atmospheric motion*, Dover Books on Earth Sciences, Dover Publications, Ney York, 1995, formerly: The Ceaseless Wind, unabridged and unaltered republication.

135. G. Dziuk, *Finite elements for the Beltrami operator on arbitrary surfaces*, Partial Differential Equations and Calculus of Variations (Berlin, Heidelberg, New York) (S. Hildebrand and R. Leis, eds.), Lecture Notes in Mathematics, vol. 1357, Springer Verlag, 1988, pp. 142–155.

136. S. Edouard, B. Legras, F. Lefèvre, and R. Eymard, *The effect of small-scale inhomogeneities on ozone depletion in the arctic*, Nature **384** (1996), 444–447.

137. H. Elbern, H. Schmidt, O. Talagrand, and A. Ebel, *4D-variational data assimilation with an adjoint air quality model for emission analysis*, Environ. Modelling & Software **15** (2000), 539–548.

138. H. Engels, *Numerical quadrature and cubature*, Computational Mathematics and Applications, Academic Press, London, 1980.

139. K. Eriksson and C. Johnson, *Adaptive finite element methods for parabolic problems I: A linear model problem*, SIAM J. Numer. Anal. **28** (1991), no. 1, 43–77.

140. C. Eskilsson and S. J. Sherwin, *A triangular spectral/hp discontinuous Galerkin method for modelling 2D shallow water equations*, Int. J. Numer. Meth. Fluids **45** (2004), 605–623.

141. R. E. Ewing and H. Wang, *A summary of numerical methods for time-dependent advection-dominated partial differential equations*, J. Comput. Appl. Math. **128** (2001), 423–445.

142. M. Falcone and R. Ferretti, *Convergence analysis for a class of high-order semi-Lagrangian advection schemes*, SIAM J. Numer. Anal. **35** (1998), no. 3, 909–940.

143. _____, *Semi-Lagrangian schemes for Hamilton-Jacobi equations, discrete representation formulae and Godunov methods*, J. Comput. Phys. **175** (2002), 559–575.

144. C. C. Fang and T. W. H. Sheu, *Two element-by-element iterative solutions for shallow water equations*, SIAM J. Sci. Comput. **22** (2001), no. 6, 2075–2092.

145. M. Farhloul and M. Fortin, *Review and complements on mixed-hybrid finite element methods for fluid flows*, J. Comput. Appl. Math. **140** (2002), 301–313.

146. B. H. Fiedler, *Grid adaption and its effect on entrainment in an E-l model of the atmospheric boundary layer*, Mon. Wea. Rev. **130** (2002), 733–740.

147. M. Fiedler, *Algebraic connectivity of graphs*, Czechoslovak Mathematical Journal **23** (1973), 298–305.

148. _____, *A property of eigenvectors of nonnegative symmetric matrices and its application to graph theory*, Czechoslovak Mathematical Journal **25** (1975), 619–633.

149. G. J. Fix, M. D. Gunzburger, and R. A. Nicolaides, *On finite element methods of the least squares type*, Comp. Math. with Appls. **5** (1979), 87–98.

150. J. E. Flaherty, R. M. Loy, M. S. Shephard, B. K. Szymanski, J. D. Teresco, and L. H. Ziantz, *Adaptive local refinement with octree load balancing for the parallel solution of three-dimensional conservation laws*, J. Par. Dist. Comp. **47** (1997), 139–152.

151. R. Ford, C. C. Pain, M. D. Piggott, A. H. J. Goddard, C. R. E. de Oliveira, and A. P. Umpleby, *A nonhydrostatic finite-element model for three-dimensional stratified oceanic flows. Part I: Model formulation*, Mon. Wea. Rev. **132** (2004), 2816–2831.

152. R. Ford, C. C. Pain, M. D. Piggott, A. J. H. Goddard, C. R. E. de Oliveira, and A. P. Umpleby, *A nonhydrostatic finite-element model for three-dimensional stratified flows. Part II: Model validation*, Mon. Wea. Rev. **132** (2004), 2832–2844.

153. B. Fornberg and N. Flyer, *Accuracy of radial basis function interpolation and derivative approximations on 1-D infinite grids*, http://amath.colorado.edu/faculty/fornberg/Docs/RBF.pdf, 2003.

154. A. Fournier, M. A. Taylor, and J. J. Tribbia, *The spectral element atmosphere model (SEAM): High-resolution parallel computation and localized resolution of regional dynamics*, Mon. Wea. Rev. **132** (2004), 726–748.

155. J. Frank and S. Reich, *Conservation properties of smoothed particle hydrodynamics applied to the shallow water equations*, BIT Num. Math. **43** (2003), 41–55.

156. S. Frickenhaus, W. Hiller, and M. Best, *FoSSI: The family of simplified solver interfaces for the rapid development of parallel numerical atmosphere and ocean models*, Ocean Modelling **10** (2005), 185–191.

157. L. M. Frohn, J. H. Christensen, and J. Brandt, *Development of a high-resolution nested air pollution model – the numerical approach*, J. Comput. Phys. **179** (2002), 68–94.

158. J. Fuhrmann and H. Langmach, *Stability and existence of solutions of time-implicit finite volume schemes for viscous nonlinear conservation laws*, Appl. Num. Math. **37** (2001), 201–230.

159. J. Galewsky, R. K. Scott, and L. M. Polvani, *An initial-value problem for testing numerical models of shallow water equations*, Tellus **56A** (2004), 429–440.

160. P. Garcia-Navarro, M. E. Hubbard, and A. Priestley, *Genuinely multidimensional upwinding for the 2D shallow water equations*, J. Comput. Phys. **121** (1995), 79–93.

161. A. George and J. W. H. Liu, *A fast implementation of the minimum degree algorithm using quotient graphs*, ACM Trans. on Math. Software **6** (1980), no. 3, 337–358.

162. P. Geuzaine, C. Grandmont, and C. Farhat, *Design and analysis of ALE schemes with provable second-order time-accuracy for inviscid and viscous flow simulations*, J. Comput. Phys. **191** (2003), 206–227.

163. S. Ghorai, A. S. Tomlin, and M. Berzins, *Resolution of pollutant concentrations using a fully 3D adaptive method*, Atmospheric Modelling: IMA Volumes in Mathematics and Applications (New York) (Chock and Charmichael, eds.), Springer Verlag, 2002, pp. 61–79.

164. R. Giering and T. Kaminski, *Recipes for adjoint code construction*, ACM Trans. Math. Software **24** (1998), no. 4, 437–474.

165. A. E. Gill, *Atmosphere-ocean dynamics*, International Geophysics Series, vol. 30, Academic Press, London, 1982.

166. F. X. Giraldo, *The Lagrange-Galerkin method for the two-dimensional shallow water equations on adaptive grids*, Int. J. Numer. Meth. Fluids **33** (2000), 789–832.

167. _____, *A spectral element shallow water model on spherical geodesic grids*, Int. J. Numer. Meth. Fluids **35** (2001), 869–901.

168. _____, *Strong and weak Lagrange-Galerkin spectral element methods for the shallow water equations*, Comp. Math. Appl. **45** (2003), 97–121.

169. F. X. Giraldo, J. S. Hesthaven, and T. Warburton, *Nodal high-order discontinuous Galerkin methods for the spherical shallow water equations*, J. Comput. Physics **181** (2002), 499–525.

170. F. X. Giraldo and B. Neta, *Stability analysis for Eulerian and semi-Lagrangian finite-element formulation of the advection-diffusion equation*, Comp. Math. Appl. **38** (1999), 97–112.

171. F. X. Giraldo and T. Warburton, *A nodal triangle-based spectral element method for the shallow water equations on the sphere*, J. Comput. Phys. (2005), in press.

172. P. Glaister, *Conservative upwind difference schemes for the shallow water equations*, Comp. Math. Appl. **39** (2000), 189–199.

173. G. Globisch, *PARMESH – A parallel mesh generator*, Parallel Computing **21** (1995), 509–524.

174. S. G. Gopalakrishnan, D. P. Bacon, N. N. Ahmad, Z. Boybeyi, T. J. Dunn, M. S. Hall, Y. Jin, P. C. S. Lee, D. E. Mays, R. V. Madala, A. Sarma, M. D. Turner, and T. R. Wait, *An operational multiscale hurricane forecasting system*, Mon. Wea. Rev. **130** (2002), 1830–1847.

175. D. Gottlieb and J. S. Hesthaven, *Spectral methods for hyperbolic problems*, J. Comp. Appl. Math. **128** (2001), 83–131.

176. A. Graff and W. Joppich, *Parallelisierung eines Helmholtzlösers aus den lokalen Vorhersagemodellen des Deutschen Wetterdienstes*, Arbeitspapiere der GMD 896, Gesellschaft für Mathematik und Datenverarbeitung, Sankt Augustin, 1995.

177. S. Gravel and A. Staniforth, *A mass-conserving semi-Lagrangian scheme for the shallow-water equations*, Mon. Wea. Rev. **122** (1994), 243–248.

178. M. Griebel and G. Zumbusch, *Hash-storage techniques for adaptive multilevel solvers and their domain decomposition parallelization*, Contemporary Mathematics **218** (1998), 279–286.

179. _____, *Parallel multigrid in an adaptive PDE solver based on hashing and space-filling curves*, Parallel Computing **25** (1999), 827–843.

180. A. Griewank, *Evaluating derivatives: Principles and techniques of algorithmic differentiation*, Frontiers in Applied Mathematics, vol. 19, SIAM, Philadelphia, 2000.

181. J. J. Hack, J. M. Rosinski, D. L. Williamson, B. A. Boville, and J. E. Truesdale, *Computational design of the NCAR community climate model*, Parallel Computing **21** (1995), 1545–1569.

182. E. Hairer, S. P. Nørsett, and G. Wanner, *Solving ordinary differential equations I: Nonstiff problems*, 2nd ed., Springer Verlag, Berlin, Heidelberg, New York, 2000, 2nd corrected printing.

183. E. Hairer and G. Wanner, *Solving ordinary differential equations II: Stiff and differential-algebraic problems*, 2nd ed., Springer Verlag, Berlin, Heidelberg, New York, 2002, 2nd corrected printing.

184. P. Hall and A. M. Davies, *The influence of an irregular grid upon internal wave propagation*, Ocean Modelling **10** (2005), 193–209.

185. D. A. Ham, J. Pietrzak, and G. S. Stelling, *A scalable unstructured grid 3-dimensional finite volume model for the shallow water equations*, Ocean Modelling **10** (2005), 153–169.

186. S. W. Hammond, R. D. Loft, J. M. Dennis, and R. K. Sato, *Implementation and performance issues of a massively parallel atmospheric model*, Parallel Computing **21** (1995), 1593–1619.

187. E. Hanert, D. Y. Le Roux, V. Legat, and E. Deleersnijder, *An efficient Eulerian finite element method for the shallow water equations*, Ocean Modelling **10** (2005), 115–136.

188. E. Hanert, V. Legat, and E. Deleersnijder, *A comparison of three finite elements to solve the linear shallow water equations*, Ocean Modelling **5** (2002), 17–35.

189. R. L. Hardy, *Multiquadric equations of topography and other irregular surfaces*, J. Geophys. Res. **76** (1971), no. 8, 1905–1915.

190. Y. Hasbani, E. Livne, and M. Bercovier, *Finite elements and characteristics applied to advection-diffusion equations*, Computers and Fluids **11** (1983), no. 2, 71–83.

191. P. Haynes and J. Anglade, *The vertical-scale cascade in atmospheric tracers due to large-scale differential advection*, Jou. Atm. Sci. **54** (1997), no. 9, 1121–1136.

192. R. W. Healy and T. F. Russell, *Solution of the advection-dispersion equation in two dimensions by a finite-volume eulerian-lagrangian localized adjoint method*, Adv. Water Resources **21** (1998), no. 1, 11–26.

193. T. Heinze, *Ein numerisches Verfahren zur Lösung der Flachwassergleichungen auf einer rotierenden Kugel mittels der Lagrange-Galerkin-Methode*, Diplomarbeit, Institut für angewandte Mathematik, Meteorologisches Institut, Rheinische Friedrich-Wilhelms-Universität, Bonn, Germany, 1998.

194. T. Heinze and A. Hense, *The shallow water equations on the sphere and their Lagrange-Galerkin-solution*, Meteorol. Atmos. Phys. **81** (2002), 129–137.

195. I. M. Held and M. J. Suarez, *A proposal for the intercomparison of the dynamical cores of atmospheric general circulation models*, Bulletin of the American Meteorol. Soc. **75** (1994), no. 10, 1825–1830.

196. D. Hempel, *Local mesh adaptation in two space dimensions*, IMPACT of Comp. Sci. Engrg. **5** (1993), 309–317.

197. B. Hendrickson, *Load balancing fictions, falsehoods and fallacies*, Applied Mathematical Modelling **25** (2000), 99–108.

198. B. Hendrickson and T. G. Kolda, *Graph partitioning models for parallel computing*, Parallel Computing **26** (2000), 1519–1534.

199. B. Hendrickson and R. Leland, *The Chaco user's guide: Version 2.0*, Technical Report SAND94-2692, Sandia National Laboratory, 1994, ftp://ftp.cs.sandia.gov/pub/papers/bahendr/guide.ps.gz.

200. ———, *A multilevel algorithm for partitioning graphs*, Proceedings of the IEEE/ACM SC95 Conference, Dec. 03-06, 1995, San Diego, California, 1995.

201. R. Hess, *Dynamically adaptive multigrid on parallel computers for a semi-implicit discretization of the shallow water equations*, Tech. Report 9, GMD – Forschungszentrum Informationstechnik GmbH, St. Augustin, 1999.

202. H. A. Hijkstra, H. Oksuzoglu, F. W. Wubs, and E. F. F. Botta, *A fully implicit model of the three-dimensional thermohaline ocean circulation*, J. Comput. Phys. **173** (2001), 685–715.

203. D. Hilbert, *Über die stetige Abbildung einer Linie auf ein Flächenstück*, Math. Ann. **38** (1891), no. 3, 459–460.

204. K. Ho-Le, *Finite element mesh generation methods: A review and classification*, Computer-Aided Design **20** (1988), no. 1, 27–38.

205. R. W. Hockney and C. R. Jesshope, *Parallel computers 2: Architecture, programming, and algorithms*, 2nd ed., Adam Hilger, Bristol, Philadelphia, 1988.

206. P. Houston and E. Süli, *A posteriori error indicators for hyperbolic problems*, http://web.comlab.ox.ac.uk/oucl/publications/natr/na-97-14.html, 1998.

207. M. E. Hubbart and N. Nikiforakis, *A three-dimensional, adaptive, Godunov-type model for global atmospheric flows*, Mon. Wea. Rev. **131** (2003), 1848–1864.

208. S. Hubbert, *Radial basis function interpolation on the sphere*, Phd thesis, Imperial College London, London, U.K., 2002, http://www.math.uni-giessen.de/Numerik/hubbert.html.

209. S. Hubbert and T. M. Morton, L_p-*error estimates for radial basis function interpolation on the sphere*, J. Approx. Theory **129** (2004), 58–77.

210. J. Hugger, *A theory for local, a posteriori, pointwise, residual-based estimation of the finite element error*, J. Comput. Appl. Math. **135** (2001), 241–292.

211. W. Hundsdorfer and J. Jaffré, *Implicit-explicit time stepping with spatial discontinuous finite elements*, App. Num. Math. **45** (2003), 231–254.

212. W. Hundsdorfer, B. Koren, M. van Loon, and J. G. Verwer, *A positive finite-difference advection scheme*, J. Comput. Phys. **117** (1995), 35–46.

213. J. Hungershöfer and J.-M. Wierum, *On the quality of partitions based on space-filling curves*, Computational Science - ICCS 2002: International Conference, Amsterdam, The Netherlands, April 21-24, 2002. Proceedings, Part III (Berlin Heidelberg) (P. M. A. Sloot, C. J. K. Tan, J. J. Dongarra, and A. G. Hoekstra, eds.), Lecture Notes in Computer Science, vol. 2331, Springer Verlag, 2002, pp. 36–45.

214. J. P. Iselin, W. J. Gutowski, and J. M. Prusa, *Tracer advection using dynamic grid adaptation and MM5*, Mon. Wea. Rev. **133** (2005), 175–187.

215. J. P. Iselin, J. M. Prusa, and W. J. Gutowski, *Dynamic grid adaptation using the MPDATA scheme*, Mon. Wea. Rev. **130** (2002), 1026–1039.

216. M. Iskandarani, J. C. Lewin, B.-J. Choi, and D. B. Haidvogel, *Comparison of advection schemes for high-order h-p finite element and finite volume methods*, Ocean Modelling **10** (2005), 233–252.

217. A. Iske, *Multiresolution methods in scattered data modelling*, Lecture Notes in Computational Science and Engineering, vol. 37, Springer Verlag, Berlin, Heidelberg, 2004.

218. A. Iske and M. Käser, *Conservative semi-Lagrangian advection on adaptive unstructured meshes*, Report TUM-M0207, TU München, München, 2003.

219. M. Israeli, N. H. Naik, and M. A. Cane, *An unconditionally stable scheme for the shallow water equations*, Mon. Wea. Rev. **128** (2000), 810–823.

220. S. A. Ivanenko and G. V. Muratova, *Adaptive grid shallow water modeling*, Appl. Num. Math. **32** (2000), 447–482.

221. C. Jablonowski, *Adaptive grids in weather and climate modeling*, PhD thesis, The University of Michigan, Ann Arbor, 2004, http://www.scd.ucar.edu/css/staff/cjablono/amr.html.

222. R. Jakob-Chien, J. J. Hack, and D. L. Williamson, *Spectral transform solutions to the shallow water test set*, Jour. Comp. Phys. **119** (1995), 164–187.

223. L. Jameson and T. Miyama, *Wavelet analysis and ocean modeling: A dynamically adaptive numerical method "WOFD-AHO"*, Mon. Wea. Rev. **128** (2000), 1536–1548.

224. P. K. Jimack, *An overview of parallel dynamic load-balancing for parallel adaptive computational mechanics codes*, Parallel and Distributed Processing for Computational Mechanics: Systems and Tools (B. H. V. Topping, ed.), Saxe-Coburg Publications, 1999, pp. 350–369.

225. ———, *Techniques for parallel adaptivity*, High Performance Computing for Computational Mechanics (B. H. V. Topping and L. Lammer, eds.), Saxe-Coburg Publications, 2000, pp. 105–118.

226. C. Johnson, R. Rannacher, and M. Boman, *Numerics and hydrodynamic stability: Toward error control in computational fluid dynamics*, SIAM J. Numer. Anal. **32** (1995), no. 4, 1058–1079.

227. M. T. Jones and P. E. Plassmann, *Parallel algorithms for adaptive mesh refinement*, SIAM J. Sci. Comput. **18** (1997), no. 3, 686–708.

228. A. Kageyama and T. Sato, *The "Yin-Yang Grid": An overset grid in spherical geometry*, Preprint, Earth Simulator Center, Japan Agency for Marine-Earth Science and Technology, Yokohama 236-0001, Japan, 2004, arXiv:physics/0403123.

229. B. K. Karamete, M. W. Beall, and M. S. Shephard, *Triangulation of arbitrary polyhedra to support automatic mesh generators*, Int. J. Numer. Meth. Engng. **49** (2000), 167–191.

230. S. Karni, A. Kurganov, and G. Petrova, *A smoothness indicator for adaptive algorithms for hyperbolic systems*, J. Comput. Phys. **178** (2002), 323–341.

231. G. Karypis and V. Kumar, *Metis – a software package for partitioning unstructured graphs, partitioning meshes, and computing fill-reducing orderings of sparse matrices*, University of Minesota, Dept. of Computer Science/ Army HPC Research Center, Mineapolis, MN 55455, 1998, Version 4.0.

232. ———, *Multilevel k-way partitioning scheme for irregular graphs*, J. Par. Distr. Comp. **48** (1998), 96–129.

233. ———, *A parallel algorithm for multilevel graph partitioning and sparse matrix ordering*, J. Par. Distr. Comp. **48** (1998), 71–95.

234. G. Karypis, K. Schloegel, and V. Kumar, *ParMetis parallel graph partitioning and sparse matrix ordering library – version 3.1*, Manual, University of Minnesota, Dept. of Computer Science and Engineering, Army HPC Research Center, Minneapolis, MN 55455, 2003,
http://www-users.cs.umn.edu/~karypis/metis/parmetis/files/manual.pdf.

235. K. Kashiyama, H. Ito, M. Behr, and T. E. Tezduyar, *Three-step explicit finite element computation of shallow water flows on a massively parallel computer*, Int. Jou. Num. Meth. Fluids **21** (1995), 885–900.

236. K. Kashiyama and T. Okada, *Automatic mesh generation method for shallow water flow analysis*, Int. J. Numer. Meth. Fluids **15** (1992), 1037–1057.

237. K. Kashiyama, K. Saitoh, M. Behr, and T. E. Tezduyar, *Parallel finite element methods for large-scale computation of storm surges and tidal flows*, Int. Jou. Num. Meth. Fluids **24** (1997), 1371–1389.

238. B. W. Kernighan and S. Lin, *An efficient heuristic procedure for partitioning graphs*, Bell Syst. Tech. J. **49** (1970), 291–307.

239. H.-P. Kersken, B. Fritzsch, O. Schenk, W. Hiller, J. Behrens, and E. Krauße, *Parallelization of large scale ocean models by data decomposition*, High-Performance Computing and Networking (Berlin) (W. Gentzsch and U. Harms, eds.), Lecture Notes in Computer Science, no. 796, Springer-Verlag, 1994, pp. 323–328.

240. M. Kessler, *Development and analysis of an adaptive transport scheme*, Atmospheric Environment **33** (1999), 2347–2360.

241. R. M. Kirby and G. E. Karniadakis, *De-aliasing on non-uniform grids: algorithms and applications*, J. Comput. Phys. **191** (2003), 249–264.

242. L. Klassen, D. Kröner, and Ph. Schott, *Finite volume method on unstructured grids in 3D with applications to the simulation of gravity waves*, Meteorol. Atmos. Phys. **82** (2003), 259–270.

243. R. Klöfkorn, D. Kröner, and M. Ohlberber, *Local adaptive methods for convection dominated problems*, Int. J. Numer. Meth. Fluids **40** (2002), 79–91.

244. I. M. Klucewicz, *A piecewise C^1 interpolant to arbitrarily spaced data*, Comp. Graph. and Image Proc. **8** (1978), 92–112.

245. I. Knowles and R. Wallace, *A variational method for numerical differentiation*, Numer. Math. **70** (1995), 91–110.

246. D. Kröner, S. Noelle, and M. Rokyta, *Convergence of higher order upwind finite volume schemes on unstructured grids for scalar conservation laws in several space dimensions*, Numer. Math. **71** (1995), 527–560.

247. Y. Kurihara and M. A. Bender, *Use of a movable nested-mesh model for tracking a small vortex*, Mon. Wea. Rev. **108** (1980), 1792–1809.

248. Y. Kurihara, G. J. Tripoli, and M. A. Bender, *Design of a movable nested-mesh primitive equation model*, Mon. Wea. Rev. **107** (1979), 239–249.

249. G. Labadie, J. P. Benque, and B. Latteux, *A finite element method for the shallow water equations*, Numerical methods in laminar and turbulent flow; Proceedings of the Second International Conference, Venice, Italy, July 13-16, 1981. (A83-23176 08-34) Swansea, Wales, Pineridge Press, 1981, p. 681-692., 1981, pp. 681–692.

250. Z. Lan, V. E. Taylor, and G. Bryan, *A novel dynamic load balancing scheme for parallel systems*, J. Parallel Distrib. Comput. **62** (2002), 1763–1781.

251. J. Lang, W. Cao, W. Huang, and R. D. Russell, *A two-dimensional moving finite element method with local refinement based on a posteriori error estimates*, App. Num. Math. **46** (2003), 75–94.

252. D. Lanser, J. G. Blom, and J. G. Verwer, *Spatial discretization of the shallow water equations in spherical geometry using Osher's scheme*, Jou. Comp. Phys. **165** (2000), 542–565.

253. _____, *Spatial discretization of the shallow water equations in spherical geometry using Osher's scheme*, J. Comput. Phys. **165** (2000), 542–565.

254. _____, *Time integration of the shallow water equations in spherical geometry*, J. Comput. Phys. **171** (2001), 373–393.

255. M. Läuter, *An adaptive Lagrange-Galerkin method for the shallow water equations on the sphere*, PAMM **3** (2003), 48–51.

256. _____, *Großräumige Zirkulationsstrukturen in einem nichtlinearen adaptiven Atmosphärenmodell*, PhD thesis, Mathematisch-Naturwissenschaftliche Fakultät der Universität Potsdam, Potsdam, Germany, 2004.

257. M. Läuter, D. Handorf, and K. Dethloff, *Unsteady analytical solutions of the spherical shallow water equations*, article in press, Alfred-Wegener-Institute for Polar and Marine Research, Potsdam, Germany, 2005.

258. M. Läuter, D. Handorf, K. Dethloff, S. Frickenhaus, N. Rakowsky, and W. Hiller, *An adaptive Lagrange-Galerkin shallow-water model on the sphere*, Proceedings of the Workshop on Current Development in Shallow Water Models on the Sphere, March 10–14, 2003, Garching, Germany (Boltzmannstr. 3, 85747 Garching, Germany) (Th. Heinze, D. Lanser, and A. T. Layton, eds.), TU München, Center for Mathematical Sciences, 2004, http://www-m3.ma.tum.de/m3/workshop/proceedings.html.

259. A. T. Layton and W. F. Spotz, *A semi-Lagrangian double Fourier method for the shallow water equations on the sphere*, J. Comput. Phys. **189** (2003), 180–196.

260. F.-X. Le Dimet and O. Talagrand, *Variational algorithms for analysis and assimilation of meteorological observations: theoretical aspects*, Tellus **38A** (1986), 97–110.

261. D. Y. Le Roux, C. A. Lin, and A. Staniforth, *A semi-implicit semi-Lagrangian finite-elment shallow-water ocean model*, Mon. Wea. Rev. **128** (2000), 1384–1401.

262. R. LeVeque, J. O. Langseth, M. Berger, and S. Mitran, *Clawpack homepage*, http://www.amath.washington.edu/~claw/.

263. R. J. LeVeque, *Numerical methods for conservation laws*, 2nd ed., Lectures in Mathematics ETH Zürich, Birkhäuser Verlag, Basel, Boston, Berlin, 1992.

264. _____, *Finite volume methods for hyperbolic problems*, Cambridge Texts in Applied Mathematics, Cambridge University Press, Cambridge, UK, 2002.

265. D. Lewis and N. Nigam, *Geometric integration on spheres and some interesting applications*, J. Comput. Appl. Math. **151** (2003), 141–170.

266. G. W. Ley and R. L. Elsberry, *Forecasts of typhoon Irma using a nested-grid model*, Mon. Wea. Rev. **104** (1976), 1154–1161.

267. P. Lin, K. W. Morton, and E. Süli, *Characteristic galerkin schemes for scalar conservation laws in two and three space dimensions*, SIAM J. Numer. Anal. **34** (1997), no. 2, 779–796.

268. S.-J. Lin, *A "Vertically Lagrangian" finite-volume dynamical core for global models*, Mon. Wea. Rev. **132** (2004), 2293–2307.

269. S.-J. Lin and R. B. Rood, *Multidimensional flux-form semi-Lagrangian transport schemes*, Mon. Wea. Rev. **124** (1996), 2046–2070.

270. X. Liu, *Four alternative patterns of the Hilbert curve*, App. Math. Comput. **147** (2004), 741–752.

271. B. Machenhauer and M. Olk, *The development of a cell-integrated semi-Lagrangian shallow water model on the sphere*, ECMWF Semi-Lagrangian Workshop 6-8 Sept. 1995, 1995.

272. L. Machiels, J. Peraire, and A. T. Patera, *A posteriori finite-element output bounds for the incompressible Navier-Stokes equations: Application to a natural convection problem*, J. Comput. Phys. **172** (2001), 401–425.

273. A. Majda, *Introduction to PDEs and waves for the atmosphere and ocean*, Courant lecture notes in mathematics, American Mathematical Society, Providence, Rhode Island, 2003.

274. A. J. Majda and R. Klein, *Systematic multiscale models for the tropics*, J. Atmos. Sci. **60** (2003), 393–408.

275. P. A. Makar and S. R. Karpik, *Basis-spline interpolation on the sphere: Applications to semi-Lagrangian advection*, Mon. Wea. Rev. **124** (1996), 182–199.

276. N. Martin and S. M. Gorelick, *Semi-analytical method for departure point determination*, Int. J. Numer. Meth. Fluids **47** (2005), 121–137.

277. The Mathworks, Inc., Natick, MA, *Partial differential equations toolbox user's guide*, 2004,
http://www.mathworks.com/access/helpdesk/help/pdf_doc/pde/pde.pdf.

278. D. J. Mavriplis, *Unstructured grid techniques*, Annu. Rev. Fluid. Mech. **29** (1997), 473–514.

279. A. McDonald, *Accuracy of multiply-upstream, semi-Lagrangian advective schemes*, Mon. Wea. Rev. **112** (1984), 1264–1275.

280. ———, *Accuracy of multiply-upstream, semi-Lagrangian advective schemes II*, Mon. Wea. Rev. **115** (1987), 1446–1450.

281. A. McDonald and J. R. Bates, *Improving the estimate of the departure point position in a two-time level semi-Lagrangian and semi-implicit scheme*, Mon. Wea. Rev. **115** (1987), 737–739.

282. A. McDonald and J. R. Bates, *Semi-Lagrangian integration of a gridpoint shallow-water model on the sphere.*, Mon. Wea. Rev. **117** (1989), 130–137.

283. A. Meister and T. Sonar, *Finite-volume schemes for compressible fluid flow*, Surv. Math. Ind. **8** (1998), 1–36.

284. J. Mellor-Crummey, D. Whalley, and K. Kennedy, *Improving memory hierarchy performance for irregular applications*, Proceedings of the 13th International Conference on Supercomputing (Rhodes, Greece), 1999, ISBN: 1-58113-164-X, pp. 425–433.

285. L. Mentrup, *Entwicklung einer massenerhaldenden Semi-Lagrange-Methode zur Simulation von Spurenstofftransport in der Atmosphäreauf einem adaptiven dreidimensionalen Gitter*, diploma thesis, TU München, Zentrum Mathematik, Garching, Germany, 2003,
http://www-m3.ma.tum.de/m3/mentrup/DA_MPSLM.pdf.

286. C. A. Micchelli, *Interpolation of scattered data: Distance matrices and conditionally positive definite functions*, Constr. Approx. **2** (1986), 11–22.

287. K. A. Mironakis and P. A. Kassomenos, *Application of MM5 model in the northwest area of Greece using a four-nest procedure*, Int. J. Envir. Pollut. **20** (2003), no. 1–6, 269–277.

288. A. R. Mitchell and D. F. Griffiths, *The finite difference method in partial differential equations*, John Wiley & Sons, Chichester, New York, Brisbane, Toronto, 1980.

289. S. A. Mitchell and S. A. Vavasis, *Quality mesh generation in higher dimensions*, SIAM J. Comput. **29** (2000), no. 4, 1334–1370.

290. W. F. Mitchell, *A comparison of adaptive refinement techniques for elliptic problems*, ACM Trans. in Math. Softw. **15** (1989), no. 4, 326–347.

291. K. W. Morton, *Discretization of unsteady hyperbolic conservation laws*, SIAM J. Numer. Anal. **39** (2001), no. 5, 1556–1597.

292. K. W. Morton, A. Priestley, and E. Süli, *Stability of the Lagrange-Galerkin method with non-exact integration*, Mathematical Modelling and Numerical Analysis **22** (1988), no. 4, 625–653.

293. V. A. Mousseau, D. A. Knoll, and J. M. Reisner, *An implicit nonlinearly consistent method for the two-dimensional shallow-water equations with Coriolis force*, Mon. Wea. Rev. **130** (2002), 2611–2625.

294. R. D. Nair and B. Machenhauer, *The mass-conservative cell-integrated semi-Lagrangian advection scheme on the sphere*, Mon. Wea. Rev. **130** (2002), 649–667.

295. R. D. Nair, J. S. Scroggs, and F. H. M. Semazzi, *Efficient conservative global transport schemes for climate and atmospheric chemistry models*, Mon. Wea. Rev. **130** (2002), 2059–2073.

296. R. D. Nair, S. J. Thomas, and R. D. Loft, *A discontinuous Galerkin global shallow water model*, Mon. Wea. Rev. **133** (2005), 876–888.

297. ———, *A discontinuous Galerkin transport scheme on the cubed sphere*, Mon. Wea. Rev. **133** (2005), 814–828.

298. T. Nakamura, R. Tanaka, T. Yabe, and K. Takizawa, *Exactly conservative semi-Lagrangian scheme for multi-dimensional hyperbolic equations with directional splitting technique*, J. Comput. Phys. **174** (2001), 171–207.

299. F. J. Narcowich and J. D. Ward, *Scattered data interpolation of spheres: Error estimates and locally supported basis functions*, SIAM J. Math. Anal. **33** (2002), no. 6, 1393–1410.

300. B. Neta and R. T. Williams, *Stability and phase speed for various finite element formulations of the advection equation*, Computers and Fluids **14** (1986), no. 4, 393–410.

301. J. Nordström, K. Forsberg, C. Adamsson, and P. Eliasson, *Finite volume methods, unstructured meshes and strict stability for hyperbolic problems*, App. Numer. Math. **45** (2003), 453–473.

302. J. M. Oberhuber and K. Ketelsen, *Parallelization of an OCGM on the Cray T3D*, personal communication, 1994.

303. J. T. Oden, *The best FEM*, Finite Elements in Analysis and Design **7** (1990), 103–114.

304. R. Oehmke and Q. F. Stout, *Parallel adaptive blocks on a sphere*, Proc. 11th SIAM Conf. Parallel Processing for Sci. Computing, 2001,
http://www.eecs.umich.edu/~qstout/pap/SIAMPP01.ps.

305. C. Ollivier-Gooch and M. VanAltena, *A high-order-accurate unstructured mesh finite-volume scheme for the advection-diffusion equation*, J. Comput. Phys. **181** (2002), 729–752.

306. S. J. Owen, *CUBIT homepage*,
http://cubit.sandia.gov.

307. ———, *Meshing research corner homepage*,
http://www.andrew.cmu.edu/user/sowen/mesh.html.

308. C. C. Pain, M. D. Piggott, A. J. H. Goddard, F. Fang, G. J. Gorman, D. P. Marshall, M. D. Eaton, P. W. Power, and C. R. E. de Oliveira, *Three-dimensional unstructured mesh ocean modelling*, Ocean Modelling **10** (2005), 5–33.

309. R. Pasquetty and F. Rapetti, *Spectral element methods on triangles and quadrilaterals: comparison and applications*, J. Comput. Phys. **198** (2004), 349–362.

310. G. Peano, *Sur une courbe, qui remplit toute une aire plane*, Math. Ann. **36** (1890), no. 1, 157–160.

311. J. Pedlosky, *Geophysical fluid dynamics*, 2nd ed., Springer-Verlag, New York, 1987.

312. F. Pellegrini, *SCOTCH 3.4 user's guide*, Research Report RR-1264-01, Laboratoire Bordelais de Recherche en Informatique, Université Bordeaux I, Bordeaux, France, 2001,
http://www.labri.fr/Perso/~pelegrin/papers/scotch_user3.4.ps.gz.

313. X. Peng, F. Xiao, T. Yabe, and K. Tani, *Implementation of the CIP as the advection solver in the MM5*, Mon. Wea. Rev. **131** (2003), 1256–1271.

314. D. W. Pepper and D. B. Carrington, *Application of h-adaptation for environmental fluid flow and species transport*, Int. J. Numer. Meth. Fluids **31** (1999), no. 1, 275–283.

315. A. F. Pereira, *Numerical investigation of tidal processes and phenomena in the Wedell Sea, Antarctica*, PhD thesis, Universität Bremen, Bremen, Germany, 2001,
http://elib.suub.uni-bremen.de/publications/dissertations/E-Diss233_cover.pdf.

316. L. Pesch, *A finite-volume discretization of the shallow-water equations in spherical geometry*, Proceedings of the Workshop on Current Development in Shallow Water Models on the Sphere, March 10–14, 2003, Garching, Germany (Boltzmannstr. 3, 85747 Garching, Germany) (Th. Heinze, D. Lanser, and A. T. Layton, eds.), TU München, Center for Mathematical Sciences, 2004,
http://www-m3.ma.tum.de/m3/workshop/proceedings.html.

317. N. A. Phillips, *The general circulation of the atmosphere: A numerical experiment*, Quart. J. Roy. Meteor. Soc. **82** (1956), 123–164.

318. _____, *A coordinate system having some special advantages for numerical forecasting*, Journal of Atmospheric Sciences **14** (1957), no. 2, 184–185.

319. _____, *A map projection system suitable for large-scale numerical weather prediction*, J. Meteor. Soc. Japan (1957), 262–267.

320. T. N. Phillips and A. J. Williams, *Conservative semi-Lagrangian finite volume schemes*, Numer. Methods Partial Differential Eq. **17** (2001), no. 4, 403–425.

321. M. D. Piggott, C. C. Pain, G. J. Gorman, P. W. Power, and A. H. J. Goddard, *h, r, and hr adaptivity with applications in numerical ocean modelling*, Ocean Modelling **10** (2005), 95–113.

322. J. R. Pilkington and S. B. Baden, *Dynamic partitioning of non-uniform structured workloads with spacefilling curves*, IEEE Trans. Par. Distr. Systems **7** (1996), no. 3, 288–300.

323. R. A. Plumb, D. W. Waugh, R. J. Atkinson, P. A. Newman, M. R. Lait, M. R. Schoeberl, E. V. Browell, A. J. Simmons, and M. Loewenstein, *Intrusions into the lower stratospheric Arctic vortex during the winter of 1991-1992*, J. Geophys. Res. **99** (1994), 1089–1105.

324. L. M. Polvani, R. K. Scott, and S. J. Thomas, *Numerically converged solutions of the global primitive equations for testing the dynamical core of atmospheric GCMs*, Mon. Wea. Rev. **132** (2004), 2539–2552.

325. A. Pothen, H. D. Simon, and K.-P. Liou, *Partitioning sparse matrices with eigenvectors of graphs*, SIAM J. Matrix Anal. Appl. **11** (1990), no. 3, 430–452.

326. R. Preis and R. Diekmann, *The PARTY partitioning-library, user guide – version 1.1*, Technical Report TR-RSFB-96-024, University of Paderborn, Paderborn, Germany, 1996,
ftp://ftp.uni-paderborn.de/doc/techreports/Informatik/tr-rsfb-96-024.ps.Z.

327. A. Priestley, *A quasi-conservative version of the semi-Lagrangian advection scheme*, Mon. Wea. Rev. **121** (1993), 621–629.

328. _____, *Exact projections and Lagrange-Galerkin method: A realistic alternative to quadrature*, J. Comp. Phys. **112** (1994), no. 2, 316–333.

329. _____, *The positive and nearly conservative Lagrange-Galerkin method*, IMA Journal of Numerical Analysis **14** (1994), 277–294.

330. J. M. Prusa and P. K. Smolarkiewicz, *An all-scale anelastic model for geophysical flows: dynamic grid deformation*, J. Comput. Phys. **190** (2003), 601–622.

331. J. A. Pudykiewicz, *Application of adjoint tracer transport equations for evaluating source parameters*, Atmospheric Environment **32** (1998), no. 17, 3039–3050.

332. N. Rakowsky, S. Frickenhaus, W. Hiller, M. Läuter, D. Handorf, and K. Dethloff, *A self-adaptive finite element model of the atmosphere*, ECMWF Workshop on the Use of High Performance Computing in Meteorology: Realizing Tera Computing, 4–8 November, Reading, UK (Singapore) (W. Zwieflhofer and N. Kreitz, eds.), ECMWF, World Scientific, 2003, pp. 279–293.

333. R. Redler, K. Ketelsen, J. Dengg, and C. W. Böning, *A high-resolution numerical model for the circulation of the Atlantic ocean*, Contribution to the 4th CRAY-SGI MPP Workshop, Garching/Munich, Sept. 10-12, 1998, 1998.

334. W. H. Reed and T. R. Hill, *Triangular mesh methods for the neutron transport equation*, report LA-UR-73-479, Los Alamos Nat. Lab., Los Alamos, NM, USA, 1973.

335. J. Reisner, V. Mousseau, and D. Knoll, *Application of the Newton-Krylov method to geophysical flows*, Mon. Wea. Rev. **129** (2001), 2404–2415.

336. J.-F. Remacle, J. E. Flaherty, and M. S. Shephard, *An adaptive discontinuous Galerkin technique with an orthogonal basis applied to compressible flow problems*, SIAM Review **45** (2003), no. 1, 53–72.

337. R. J. Renka, *Algorithm 624: Triangulation and interpolation at arbitrarily distributed points in the plane*, ACM Trans. on Math. Softw. **10** (1984), no. 4, 440–442.

338. _____, *Algorithm 661 QSHEP3D: Quadratic Shepard method for trivariate interpolation of scattered data*, ACM Trans. Math. Softw. **14** (1988), no. 2, 151–152.

339. T. D. Ringler and D. A. Randall, *A potential enstrophy and energy conserving numerical scheme for solution of the shallow-water equations on a geodesic grid*, Mon. Wea. Rev. **130** (2002), 1397–1410.

340. A. Rinke, K. Dethloff, and J. H. Christensen, *Arctic winter climate and its interannual variation simulated by a regional climate model*, J. Geophys. Res. **104** (1999), 19,027–19,038.

341. M. C. Rivara, *Algorithms for refining triangular grids suitable for adaptive and multigrid techniques*, International Journal for Numerical Methods in Engineering **20** (1984), 745–756.

342. L. Rivier, R. Loft, and L. M. Polvani, *An efficient spectral dynamical core for distributed memory computers*, Mon. Wea. Rev. **130** (2002), 1384–1396.

343. A. Robert, *A stable numerical integration scheme for the primitive meteorological equations*, Atmosphere-Ocean **19** (1981), 35–46.

344. S. Roberts, S. Kalyanasundaram, M. Cardew-Hall, and W. Clarke, *A key based parallel adaptive refinement technique for finite element methods*, Tech. report, Australian National University, Canberra, ACT 0200, Australia, 1997.

345. C. Ronchi, R. Iacono, and P. S. Paolucci, *The "cubed sphere": A new method for the solution of partial differential equations in spherical geometry*, J Comput. Phys. **124** (1996), 93–114.

346. R. Rosen, *Matrix bandwidth minimization*, ACM/CSC-ER Proceedings of the 1968 23rd ACM national conference (New York, NY, USA), ACM Press, 1968, pp. 585–595.

347. D. Rosenberg, A. Fournier, P. Fischer, and A. Pouquet, *Geophysical-astrophysical spectral-element adaptive refinement (GASpAR): Object-oriented h-adaptive fluid dynamics simulation*, J. Comput. Phys. **215** (2006), 59–80.

348. G. Roussos and B. J. C. Baxter, *Rapid evaluation of radial basis functions*, J. Comput. Appl. Math. **180** (2005), 51–70.

349. T. F. Russell and R. V. Trujillo, *Eulerian-Lagrangian localized adjoint methods with variable coefficients in multiple dimensions*, Computational Methods in Surface Hydrology – Proceedings of the Eighth International Conference on Computational Methods in Water Resources, held in Venice, Italy, June 11-15 1990 (Berlin) (G. Gambolati, A. Rinaldo, C. A. Brebbia, W. G. Gray, and G. F. Pinder, eds.), Springer Verlag, 1990, pp. 357–363.

350. Robert Sadourny, *The dynamics of finite-difference models of the shallow-water equations*, J. Atmos. Sci. **32** (1975), 680–689.

351. J. S. Sawyer, *A semi-Lagrangian method of solving the vorticity advection equation*, Tellus **15** (1963), 336–342.

352. K. Schloegel, G. Karypis, and V. Kumar, *Multilevel diffusion schemes for repartitioning of adaptive meshes*, J. Par. Distr. Comp. **47** (1997), 109–124.

353. A. Schmidt and K. G. Siebert, *ALBERTA homepage*, http://www.alberta-fem.de.

354. _____, *Design of adaptive finite element software: The finite element toolbox ALBERTA*, Lecture Notes in Computational Science and Engineering, vol. 42, Springer Verlag, Berlin, Heidelberg, New York, 2005.

355. R. Schneiders, *Mesh generation & grid generation on the web*, http://www-users.informatik.rwth-aachen.de/~roberts/meshgeneration.html.

356. W. Schönauer and T. Adolph, *How WE solve PDEs*, J. Comput. Appl. Math. **131** (2001), 473–492.

357. J. Schröter, U. Seiler, and M. Wenzel, *Variational assimilation of geosat data into an eddy resolving model of the Gulf Stream extension area*, J. Phys. Oceanogr. **23** (1993), 925–953.

358. C. Schwab, *p- and hp-finite element methods: Theory and applications in solid and fluid mechanics*, Clarendon Press, Oxford University Press, Oxford, New York, 1998.

359. H. R. Schwarz, *Finite element methods*, Academic Press, London, 1988.

360. J. S. Scroggs and F. H. M. Semazzi, *A conservative semi-Lagrangian method for multidimensional fluid dynamics applications*, Numerical Methods for Partial Differential Equations **11** (1995), 445–452.

361. K. R. Searle, M. P. Chipperfield, S. Bekki, and J. A. Pyle, *The impact of spatial averaging on calculated polar ozone loss – 1. model experiments*, Jou. Geophys. Res. **103** (1998), no. D19, 25,397–25,408.

362. P. Seibert and A. Frank, *Source-receptor matrix calculation with a Lagrangian particle dispersion model in backward mode*, Atmos. Chem. Phys. **4** (2004), 51–63.

363. K. Shahbazi, M. Paraschivoiu, and J. Mostaghimi, *Second order accurate volume tracking based on remapping for triangular meshes*, J. Comput. Phys. **188** (2003), 100–122.

364. H. Shan, J. P. Singh, L. Oliker, and R. Biswas, *A comparison of three programming models for adaptive applications on the Origin 2000*, J. Parallel Distr. Comput. **62** (2002), 241–266.

365. C.-W. Shu, *High-order finite difference and finite volume WENO schemes and discontinuous Galerkin methods for CFD*, Int. J. Comp. Fluid Dyn. **17** (2003), no. 2, 107–118.

366. Z. Sirkes and E. Tziperman, *Finite difference of adjoint or adjoint of finite difference?*, Mon. Wea. Rev. **125** (1997), 3373–3378.

367. W. Skamarock, J. Oliger, and R. L. Street, *Adaptive grid refinement for numerical weather prediction*, J. Comput. Phys. **80** (1989), 27–60.

368. W. C. Skamarock, *Truncation error estimates for refinement criteria in nested and adaptive models*, Mon. Wea. Rev. **117** (1989), 872–886.

369. W. C. Skamarock and J. B. Klemp, *Adaptive grid refinement for two-dimensional and three-dimensional nonhydrostatic atmospheric flow*, Mon. Wea. Rev. **121** (1993), 788–804.

370. G. D. Smith, *Numerical solution of partial differential equations: Finite difference methods*, 3rd ed., Clarendon Press, Oxford, 1993.

371. P. K. Smolarkiewicz and J. A. Pudykiewicz, *A class of semi-Lagrangian approximations for fluids*, J. Atmos. Sci. **49** (1992), no. 22, 2082–2096.

372. P. K. Smolarkiewicz and J. Szmelter, *MPDATA: An edge-based unstructured-grid formulation*, J. Comput. Phys. (2005), in press.

373. R. K. Srivastava, D. S. McRae, and M. T. Odman, *An adaptive grid algorithm for air-quality modeling*, J. Comput. Phys. **165** (2000), 437–472.

374. A. Staniforth and J. Côté, *Semi-Lagrangian integration schemes for atmospheric models - a review.*, Mon. Wea. Rev. **119** (1991), 2206–2223.

375. G. Starius, *Composite mesh difference methods for elliptic boundary value problems*, Numer. Math. **28** (1977), 243–258.

376. J. Steppeler, R. Hess, U. Schättler, and L. Bonaventura, *Review of numerical methods for nonhydrostatic weather prediction models*, Meteorol. Atmos. Phys. **82** (2003), 287–301.

377. D. E. Stevens and S. Bretherton, *A forward-in-time advection schem and adaptive multilevel flow solver for nearly incompressible atmospheric flow*, J. Comput. Phys. **129** (1996), 284–295.

378. A. Stohl, M. Hittenberger, and G. Wotawa, *Validation of the Lagrangian particle dispersion model FLEXPART against large-scale tracer experiment data*, Atmos. Eviron. **32** (1998), no. 24, 4245–4264.

379. A. H. Stroud, *Approximate calculation of multiple integrals*, Prentice Hall, Inc., Englewood Cliffs, NJ, USA, 1971.

380. E. Süli, *Convergence and nonlinear stability of the Lagrange-Galerkin method for the Navier-Stokes equations*, Numer. Math. **53** (1988), no. 4, 459–483.

381. W.-Y. Sun and M.-T. Sun, *Mass correction applied to semi-Lagrangian advection scheme*, Mon. Wea. Rev. **132** (2004), 975–984.

382. B. A. Szabo, *Some recent development in finite element analysis*, Computers and Mathematics with Applications **5** (1979), 99–115.

383. M. Tanemura, T. Ogawa, and N. Ogita, *A new algorithm for three-dimensional Voronoi tesselation*, Jou. Comp. Phys. **51** (1983), no. 2, 191–207.

384. H. Tang and T. Tang, *Adaptive mesh methods for one- and two-dimensional hyperbolic conervation laws*, SIAM J. Numer. Anal. **41** (2003), no. 2, 487–515.

385. M. Tanguay, P. Bartello, and P. Gauthier, *Four-dimensional data assimilation with a wide range of scales*, Tellus **47A** (1995), 974–997.

386. M. Tanguay and S. Polavarapu, *The adjoint of the semi-Lagrangian treatment of the passive tracer equation*, Mon. Wea. Rev. **127** (1999), 551–564.

387. M. Taylor, J. Tribbia, and M. Iskandarani, *The spectral element method for the shallow water equations on the sphere*, J. Comput. Phys. **130** (1997), 92–108.

388. C. Temperton and A. Staniforth, *An efficient two-time-level semi-Lagrangian semi-implicit integration scheme*, Quart. J. Roy. Meteor. Soc. **113** (1987), 1025–1039.

389. I. Thomas and T. Sonar, *On a second order residual estimator for numerical schemes for nonlinear hyperbolic conservation laws*, J. Comput. Phys. **171** (2001), 227–242.

390. V. Thomée, *From finite differences to finite elements A short history of numerical analysis of partial differential equations*, J. Comput. Appl. Math. **128** (2001), 1–54.

391. B. K. Thompson, J. F. ad Sony and N. P. Weatherill (eds.), *Handbook of grid generation*, CRC Press, Boca Raton, London, New York, Washington D.C., 1999.

392. J. Thuburn, *Multidimensional flux-limited advection schemes*, J. Comput. Phys. **123** (1996), 74–83.

393. A. Tomlin, M. Berzins, J. Ware, J. Smith, and M. J. Pilling, *On the use of adaptive gridding methods for modelling chemical transport from multi-scale sources*, Atmos. Environ. **31** (1997), no. 18, 2945–2959.

394. A. S. Tomlin, S. Ghorai, G. Hart, and M. Berzins, *3-D Multi-scale air pollution modelling using adaptive unstructured meshes*, Environmental Modelling and Software **15** (2000), 681–692.

395. M Torrilhon and M. Fey, *Constraint-preserving upwind methods for multidimensional advection equations*, SIAM J. Numer. Anal. **42** (2004), no. 4, 1694–1728.

396. N. Touheed and P. Jimack, *Dynamic load-balancing for adaptive PDE solvers with hierarchical refinement*, Proceedings of the Eighth SIAM Conference on Parallel Processing for Scientific Computing (M. et al. Heath, ed.), SIAM, 1997.

397. M. Turner, H. C. Clough, H. C. Martin, and L. J. Topp, *Stiffness and deflection analysis of complex structures*, J. Aeronaut. Sci. **23** (1956), no. 9, 805–823.

398. L. Umscheid Jr. and M. Sankar-Rao, *Further tests of a grid system for global numerical prediction*, Mon. Wea. Rev. **99** (1971), no. 9, 686–690.

399. S. A. Vavasis, *QMG 2.0 – overview and examples of QMG*, Cornell University, Ithaca, NY, 1999,
http://www.cs.cornell.edu/home/vavasis/qmg2.0.

400. R. Verfürth, *A posteriori error estimation and adaptive mesh-refinement techniques*, J. Comp. App. Math. **50** (1994), 67–83.

401. _____, *A review of a posteriori error estimation and adaptive mesh refinement techniques*, Wiley-Teubner, Chichester, 1996.

402. N. J. Walkington, *Convergence of the discontinuous Galerkin method for discontinuous solutions*, SIAM J. Numer. Anal. **42** (2005), no. 5, 1801–1817.

403. C. Walshaw, *The parallel JOSTLE library user guide: Version 3.0*, Manual, School of Computing and Mathematical Sciences, University of Greenwich, University of Greenwich, London, SE10 9LS, UK, 2002,
http://www.gre.ac.uk/~c.walshaw/jostle/jostleplib.pdf.

404. C. Walshaw, M. Cross, and M. G. Everett, *Parallel dynamic graph partitioning for adaptive unstructured meshes*, J. Par. Dist. Comp. **47** (1997), 102–108.

405. R. A. Walters and E. J. Barragy, *Comparison of h and p finite element solutions of the shallow water equations*, Int. J. Numer. Meth. Fluids **24** (1997), 61–79.

406. H. Wang, H. K. Dahle, R. E. Ewing, M. S. Espedal, R. C. Sharpley, and Man. S., *An ELLAM scheme for advection-diffusion equations in two dimensions*, SIAM J. Sci. Comput. **20** (1999), no. 6, 2160–2194.

407. Z. J. Wang and Y. Liu, *Spectral (finite) volume method for conservation laws on unstructured grids*, J. Comput. Phys. **179** (2002), 665–697.

408. D. W. Waugh and R. A. Plumb, *Contour advection with surgery: A technique for investigating finescale structure in tracer transport*, Jou. Atm. Sci. **51** (1994), no. 4, 530–540.

409. D. W. Waugh, R. A. Plumb, R. J. Atkinson, M. R. Schoeberl, L. R. Lait, P. A. Newman, M. Loewenstein, D. W. Toohey, L. M. Avallone, C. R. Webster, and R. D. May, *Transport out of the lower stratospheric arctic vortex by Rossby wave breaking*, J. Geophys. Res. **99** (1994), 1071–1088.

410. P. Wesseling, *Principles of computational fluid dynamics*, Springer Verlag, Berlin, Heidelberg, New York, 2001.

411. J.-M. Wierum, *Anwendung diskreter raumfüllender Kurven: Graphpartitionierung und Kontaktsuche in der Finite-Elemente-Simulation*, PhD thesis, Paderborn University, Faculty of Computer Science, Electrical Engineering and Mathematics, 2003, http://wwwcs.upb.de/pc2/papers/files/423.pdf.

412. A. Wiin-Nielsen, *On the application of trajectory methods in numerical forecasting*, Tellus **11** (1959), 180–196.

413. P. Wilders and G Fotia, *A positive spatial advection scheme on unstructured meshes for tracer transport*, J. Comput. Appl. Math. **140** (2002), 809–821.

414. D. L. Williamson and G. L. Browning, *Comparison of grids and difference approximations for numerical weather prediction over a sphere*, J. Appl. Meteor. **12** (1973), no. 2, 264–274.

415. D. L. Williamson, J. B. Drake, J. J. Hack, R. Jakob, and P. N. Swarztrauber, *A standard test set for numerical approximations to the shallow water equations in spherical geometry*, J. Comp. Phys. **102** (1992), 211–224.

416. A. M. Wissink, R. D. Hornung, S. R. Kohn, S. S. Smith, and N. Elliot, *Large scale parallel structured AMR calculations using the SAMRAI framework*, SC2001 (Denver, CO), ACM, 2001.

417. S. M. Wong, Y. C. Hon, and M. A. Golberg, *Compactly supported radial basis functions for shallow water equations*, App. Math. Comput. **127** (2002), 79–101.

418. P. H. Worley and I. T. Foster, *Parallel Spectral Transform Shallow Water Model: a runtime-tunable parallel benchmark code*, 1994 IEEE Scalable High-Performance Computing Conference (SHPCC) (Los Alamitos, CA) (J. J. Dongarra and D. W. Walker, eds.), IEEE Computer Society Press, 1994, pp. 207–214.

419. F. Xiao, *A class of single-cell high-order semi-Lagrangian advection schemes*, Mon. Wea. Rev. **128** (2000), 1165–1176.

420. Y. Xing and C.-W. Shu, *High order finite difference WENO schemes with the exact conservation property for the shallow water equations*, http://www.dam.brown.edu/scicomp/publications/Reports/Y2004/BrownSC-2004-10.pdf, 2004.

421. T. Yabe, R. Tanaka, T. Nakamura, and F. Xiao, *An exactly conservative semi-Lagrangian scheme (CIP-CSL) in one dimension*, Mon. Wea. Rev. **129** (2001), 332–344.

422. I. Yavneh and J. C. McWilliams, *Efficient multigrid solution of the shallow-water balance equations*, Tech. report, National Center for Atmospheric Research, Boulder Colorado, 1993.

423. A. Younes and P. Ackerer, *Solving the advection-diffusion equation with the Eulerian-Lagrangian localized adjoint method on unstructured meshes and non uniform time stepping*, J. Comput. Phys. (2005), in press.

424. R. Young and I. MacPhedran, *Internet finite element resources*, http://www.engr.usask.ca/~macphed/finite/fe_resources/fe_resources.html.

425. S. T. Zalesak, *Fully multidimensional flux-corrected transport algorithms for fluids*, Jou. Comput. Phys. **31** (1979), 335–362.

426. G. Zängl, *An improved method for computing horizontal diffusion in a sigma-coordinate model and its application to simulations over mountainous topography*, Mon. Wea. Rev. **130** (2002), 1423–1432.

427. Q. Zhang and C.-W. Shu, *Error estimates to smooth solutions of Runge-Kutta discontinuous Galerkin methods for scalar conservation laws*, SIAM J. Numer. Anal. **42** (2004), no. 2, 641–666.

428. O. C. Zienkiewicz and J. Z. Zhu, *A simple error estimator and adaptive procedure for practical engineering analysis*, Int. J. Numer. Meth. Eng. **24** (1987), 337–357.

429. J. Zimmermann, *Dynamische Lastverteilung bei Finite-Elemente-Methoden auf Parallelrechnern mithilfe von spacefilling curves*, Thesis, Technische Universität München, Lehrstuhl für Numerische Mathematik und Wissenschaftliches Rechnen, Boltzmannstr. 3, 85748 Garching, Germany, 2001, http://www.cip.informatik.uni-muenchen.de/žimmermc/sfc/zula/zula.html.

430. G. Zumbusch, *On the quality of space-filling curve induced partitions*, Z. Angew. Math. Mech. **81** (2001), no. S1, 25–28.

Index

Editorial Policy

1. Volumes in the following three categories will be published in LNCSE:
i)   Research monographs
ii) Lecture and seminar notes
iii) Conference proceedings

Those considering a book which might be suitable for the series are strongly advised to contact the publisher or the series editors at an early stage.

2. Categories i) and ii). These categories will be emphasized by Lecture Notes in Computational Science and Engineering. **Submissions by interdisciplinary teams of authors are encouraged.** The goal is to report new developments – quickly, informally, and in a way that will make them accessible to non-specialists. In the evaluation of submissions timeliness of the work is an important criterion. Texts should be well-rounded, well-written and reasonably self-contained. In most cases the work will contain results of others as well as those of the author(s). In each case the author(s) should provide sufficient motivation, examples, and applications. In this respect, Ph.D. theses will usually be deemed unsuitable for the Lecture Notes series. Proposals for volumes in these categories should be submitted either to one of the series editors or to Springer-Verlag, Heidelberg, and will be refereed. A provisional judgment on the acceptability of a project can be based on partial information about the work: a detailed outline describing the contents of each chapter, the estimated length, a bibliography, and one or two sample chapters – or a first draft. A final decision whether to accept will rest on an evaluation of the completed work which should include

– at least 100 pages of text;
– a table of contents;
– an informative introduction perhaps with some historical remarks which should be accessible to readers unfamiliar with the topic treated;
– a subject index.

3. Category iii). Conference proceedings will be considered for publication provided that they are both of exceptional interest and devoted to a single topic. One (or more) expert participants will act as the scientific editor(s) of the volume. They select the papers which are suitable for inclusion and have them individually refereed as for a journal. Papers not closely related to the central topic are to be excluded. Organizers should contact Lecture Notes in Computational Science and Engineering at the planning stage.

In exceptional cases some other multi-author-volumes may be considered in this category.

4. Format. Only works in English are considered. They should be submitted in camera-ready form according to Springer-Verlag's specifications.
Electronic material can be included if appropriate. Please contact the publisher.
Technical instructions and/or LaTeX macros are available via
http://www.springer.com/east/home/math/math+authors?SGWID=5-40017-6-71391-0.
The macros can also be sent on request.

General Remarks

Lecture Notes are printed by photo-offset from the master-copy delivered in camera-ready form by the authors. For this purpose Springer-Verlag provides technical instructions for the preparation of manuscripts. See also *Editorial Policy*.

Careful preparation of manuscripts will help keep production time short and ensure a satisfactory appearance of the finished book.

The following terms and conditions hold:

Categories i), ii), and iii):
Authors receive 50 free copies of their book. No royalty is paid. Commitment to publish is made by letter of intent rather than by signing a formal contract. Springer-Verlag secures the copyright for each volume.

For conference proceedings, editors receive a total of 50 free copies of their volume for distribution to the contributing authors.

All categories:
Authors are entitled to purchase further copies of their book and other Springer mathematics books for their personal use, at a discount of 33,3 % directly from Springer-Verlag.

Addresses:

Timothy J. Barth
NASA Ames Research Center
NAS Division
Moffett Field, CA 94035, USA
e-mail: barth@nas.nasa.gov

Michael Griebel
Institut für Numerische Simulation
der Universität Bonn
Wegelerstr. 6
53115 Bonn, Germany
e-mail: griebel@ins.uni-bonn.de

David E. Keyes
Department of Applied Physics
and Applied Mathematics
Columbia University
200 S. W. Mudd Building
500 W. 120th Street
New York, NY 10027, USA
e-mail: david.keyes@columbia.edu

Risto M. Nieminen
Laboratory of Physics
Helsinki University of Technology
02150 Espoo, Finland
e-mail: rni@fyslab.hut.fi

Dirk Roose
Department of Computer Science
Katholieke Universiteit Leuven
Celestijnenlaan 200A
3001 Leuven-Heverlee, Belgium
e-mail: dirk.roose@cs.kuleuven.ac.be

Tamar Schlick
Department of Chemistry
Courant Institute of Mathematical
Sciences
New York University
and Howard Hughes Medical Institute
251 Mercer Street
New York, NY 10012, USA
e-mail: schlick@nyu.edu

Mathematics Editor at Springer: Martin Peters
Springer-Verlag, Mathematics Editorial IV
Tiergartenstrasse 17
D-69121 Heidelberg, Germany
Tel.: *49 (6221) 487-8185
Fax: *49 (6221) 487-8355
e-mail: martin.peters@springer.com

Lecture Notes
in Computational Science
and Engineering

Vol. 1 D. Funaro, *Spectral Elements for Transport-Dominated Equations.* 1997. X, 211 pp. Softcover. ISBN 3-540-62649-2

Vol. 2 H. P. Langtangen, *Computational Partial Differential Equations.* Numerical Methods and Diffpack Programming. 1999. XXIII, 682 pp. Hardcover. ISBN 3-540-65274-4

Vol. 3 W. Hackbusch, G. Wittum (eds.), *Multigrid Methods V.* Proceedings of the Fifth European Multigrid Conference held in Stuttgart, Germany, October 1-4, 1996. 1998. VIII, 334 pp. Softcover. ISBN 3-540-63133-X

Vol. 4 P. Deuflhard, J. Hermans, B. Leimkuhler, A. E. Mark, S. Reich, R. D. Skeel (eds.), *Computational Molecular Dynamics: Challenges, Methods, Ideas.* Proceedings of the 2nd International Symposium on Algorithms for Macromolecular Modelling, Berlin, May 21-24, 1997. 1998. XI, 489 pp. Softcover. ISBN 3-540-63242-5

Vol. 5 D. Kröner, M. Ohlberger, C. Rohde (eds.), *An Introduction to Recent Developments in Theory and Numerics for Conservation Laws.* Proceedings of the International School on Theory and Numerics for Conservation Laws, Freiburg / Littenweiler, October 20-24, 1997. 1998. VII, 285 pp. Softcover. ISBN 3-540-65081-4

Vol. 6 S. Turek, *Efficient Solvers for Incompressible Flow Problems.* An Algorithmic and Computational Approach. 1999. XVII, 352 pp, with CD-ROM. Hardcover. ISBN 3-540-65433-X

Vol. 7 R. von Schwerin, *Multi Body System SIMulation.* Numerical Methods, Algorithms, and Software. 1999. XX, 338 pp. Softcover. ISBN 3-540-65662-6

Vol. 8 H.-J. Bungartz, F. Durst, C. Zenger (eds.), *High Performance Scientific and Engineering Computing.* Proceedings of the International FORTWIHR Conference on HPSEC, Munich, March 16-18, 1998. 1999. X, 471 pp. Softcover. ISBN 3-540-65730-4

Vol. 9 T. J. Barth, H. Deconinck (eds.), *High-Order Methods for Computational Physics.* 1999. VII, 582 pp. Hardcover. ISBN 3-540-65893-9

Vol. 10 H. P. Langtangen, A. M. Bruaset, E. Quak (eds.), *Advances in Software Tools for Scientific Computing.* 2000. X, 357 pp. Softcover. ISBN 3-540-66557-9

Vol. 11 B. Cockburn, G. E. Karniadakis, C.-W. Shu (eds.), *Discontinuous Galerkin Methods.* Theory, Computation and Applications. 2000. XI, 470 pp. Hardcover. ISBN 3-540-66787-3

Vol. 12 U. van Rienen, *Numerical Methods in Computational Electrodynamics.* Linear Systems in Practical Applications. 2000. XIII, 375 pp. Softcover. ISBN 3-540-67629-5

Vol. 13 B. Engquist, L. Johnsson, M. Hammill, F. Short (eds.), *Simulation and Visualization on the Grid.* Parallelldatorcentrum Seventh Annual Conference, Stockholm, December 1999, Proceedings. 2000. XIII, 301 pp. Softcover. ISBN 3-540-67264-8

Vol. 14 E. Dick, K. Riemslagh, J. Vierendeels (eds.), *Multigrid Methods VI.* Proceedings of the Sixth European Multigrid Conference Held in Gent, Belgium, September 27-30, 1999. 2000. IX, 293 pp. Softcover. ISBN 3-540-67157-9

Vol. 15 A. Frommer, T. Lippert, B. Medeke, K. Schilling (eds.), *Numerical Challenges in Lattice Quantum Chromodynamics.* Joint Interdisciplinary Workshop of John von Neumann Institute for Computing, Jülich and Institute of Applied Computer Science, Wuppertal University, August 1999. 2000. VIII, 184 pp. Softcover. ISBN 3-540-67732-1

Vol. 16 J. Lang, *Adaptive Multilevel Solution of Nonlinear Parabolic PDE Systems.* Theory, Algorithm, and Applications. 2001. XII, 157 pp. Softcover. ISBN 3-540-67900-6

Vol. 17 B. I. Wohlmuth, *Discretization Methods and Iterative Solvers Based on Domain Decomposition.* 2001. X, 197 pp. Softcover. ISBN 3-540-41083-X

Vol. 18   U. van Rienen, M. Günther, D. Hecht (eds.), *Scientific Computing in Electrical Engineering*. Proceedings of the 3rd International Workshop, August 20-23, 2000, Warnemünde, Germany. 2001. XII, 428 pp. Softcover. ISBN 3-540-42173-4

Vol. 19   I. Babuška, P. G. Ciarlet, T. Miyoshi (eds.), *Mathematical Modeling and Numerical Simulation in Continuum Mechanics*. Proceedings of the International Symposium on Mathematical Modeling and Numerical Simulation in Continuum Mechanics, September 29 - October 3, 2000, Yamaguchi, Japan. 2002. VIII, 301 pp. Softcover. ISBN 3-540-42399-0

Vol. 20   T. J. Barth, T. Chan, R. Haimes (eds.), *Multiscale and Multiresolution Methods*. Theory and Applications. 2002. X, 389 pp. Softcover. ISBN 3-540-42420-2

Vol. 21   M. Breuer, F. Durst, C. Zenger (eds.), *High Performance Scientific and Engineering Computing*. Proceedings of the 3rd International FORTWIHR Conference on HPSEC, Erlangen, March 12-14, 2001. 2002. XIII, 408 pp. Softcover. ISBN 3-540-42946-8

Vol. 22   K. Urban, *Wavelets in Numerical Simulation*. Problem Adapted Construction and Applications. 2002. XV, 181 pp. Softcover. ISBN 3-540-43055-5

Vol. 23   L. F. Pavarino, A. Toselli (eds.), *Recent Developments in Domain Decomposition Methods*. 2002. XII, 243 pp. Softcover. ISBN 3-540-43413-5

Vol. 24   T. Schlick, H. H. Gan (eds.), *Computational Methods for Macromolecules: Challenges and Applications*. Proceedings of the 3rd International Workshop on Algorithms for Macromolecular Modeling, New York, October 12-14, 2000. 2002. IX, 504 pp. Softcover. ISBN 3-540-43756-8

Vol. 25   T. J. Barth, H. Deconinck (eds.), *Error Estimation and Adaptive Discretization Methods in Computational Fluid Dynamics*. 2003. VII, 344 pp. Hardcover. ISBN 3-540-43758-4

Vol. 26   M. Griebel, M. A. Schweitzer (eds.), *Meshfree Methods for Partial Differential Equations*. 2003. IX, 466 pp. Softcover. ISBN 3-540-43891-2

Vol. 27   S. Müller, *Adaptive Multiscale Schemes for Conservation Laws*. 2003. XIV, 181 pp. Softcover. ISBN 3-540-44325-8

Vol. 28   C. Carstensen, S. Funken, W. Hackbusch, R. H. W. Hoppe, P. Monk (eds.), *Computational Electromagnetics*. Proceedings of the GAMM Workshop on "Computational Electromagnetics", Kiel, Germany, January 26-28, 2001. 2003. X, 209 pp. Softcover. ISBN 3-540-44392-4

Vol. 29   M. A. Schweitzer, *A Parallel Multilevel Partition of Unity Method for Elliptic Partial Differential Equations*. 2003. V, 194 pp. Softcover. ISBN 3-540-00351-7

Vol. 30   T. Biegler, O. Ghattas, M. Heinkenschloss, B. van Bloemen Waanders (eds.), *Large-Scale PDE-Constrained Optimization*. 2003. VI, 349 pp. Softcover. ISBN 3-540-05045-0

Vol. 31   M. Ainsworth, P. Davies, D. Duncan, P. Martin, B. Rynne (eds.), *Topics in Computational Wave Propagation*. Direct and Inverse Problems. 2003. VIII, 399 pp. Softcover. ISBN 3-540-00744-X

Vol. 32   H. Emmerich, B. Nestler, M. Schreckenberg (eds.), *Interface and Transport Dynamics*. Computational Modelling. 2003. XV, 432 pp. Hardcover. ISBN 3-540-40367-1

Vol. 33   H. P. Langtangen, A. Tveito (eds.), *Advanced Topics in Computational Partial Differential Equations*. Numerical Methods and Diffpack Programming. 2003. XIX, 658 pp. Softcover. ISBN 3-540-01438-1

Vol. 34   V. John, *Large Eddy Simulation of Turbulent Incompressible Flows*. Analytical and Numerical Results for a Class of LES Models. 2004. XII, 261 pp. Softcover. ISBN 3-540-40643-3

Vol. 35   E. Bänsch (ed.), *Challenges in Scientific Computing - CISC 2002*. Proceedings of the Conference *Challenges in Scientific Computing*, Berlin, October 2-5, 2002. 2003. VIII, 287 pp. Hardcover. ISBN 3-540-40887-8

Vol. 36   B. N. Khoromskij, G. Wittum, *Numerical Solution of Elliptic Differential Equations by Reduction to the Interface*. 2004. XI, 293 pp. Softcover. ISBN 3-540-20406-7

Vol. 37   A. Iske, *Multiresolution Methods in Scattered Data Modelling*. 2004. XII, 182 pp. Softcover. ISBN 3-540-20479-2

Vol. 38   S.-I. Niculescu, K. Gu (eds.), *Advances in Time-Delay Systems*. 2004. XIV, 446 pp. Softcover. ISBN 3-540-20890-9

Vol. 39 S. Attinger, P. Koumoutsakos (eds.), *Multiscale Modelling and Simulation*. 2004. VIII, 277 pp. Softcover. ISBN 3-540-21180-2

Vol. 40 R. Kornhuber, R. Hoppe, J. Périaux, O. Pironneau, O. Wildlund, J. Xu (eds.), *Domain Decomposition Methods in Science and Engineering*. 2005. XVIII, 690 pp. Softcover. ISBN 3-540-22523-4

Vol. 41 T. Plewa, T. Linde, V.G. Weirs (eds.), *Adaptive Mesh Refinement – Theory and Applications*. 2005. XIV, 552 pp. Softcover. ISBN 3-540-21147-0

Vol. 42 A. Schmidt, K.G. Siebert, *Design of Adaptive Finite Element Software*. The Finite Element Toolbox ALBERTA. 2005. XII, 322 pp. Hardcover. ISBN 3-540-22842-X

Vol. 43 M. Griebel, M.A. Schweitzer (eds.), *Meshfree Methods for Partial Differential Equations II*. 2005. XIII, 303 pp. Softcover. ISBN 3-540-23026-2

Vol. 44 B. Engquist, P. Lötstedt, O. Runborg (eds.), *Multiscale Methods in Science and Engineering*. 2005. XII, 291 pp. Softcover. ISBN 3-540-25335-1

Vol. 45 P. Benner, V. Mehrmann, D.C. Sorensen (eds.), *Dimension Reduction of Large-Scale Systems*. 2005. XII, 402 pp. Softcover. ISBN 3-540-24545-6

Vol. 46 D. Kressner (ed.), *Numerical Methods for General and Structured Eigenvalue Problems*. 2005. XIV, 258 pp. Softcover. ISBN 3-540-24546-4

Vol. 47 A. Boriçi, A. Frommer, B. Joó, A. Kennedy, B. Pendleton (eds.), *QCD and Numerical Analysis III*. 2005. XIII, 201 pp. Softcover. ISBN 3-540-21257-4

Vol. 48 F. Graziani (ed.), *Computational Methods in Transport*. 2006. VIII, 524 pp. Softcover. ISBN 3-540-28122-3

Vol. 49 B. Leimkuhler, C. Chipot, R. Elber, A. Laaksonen, A. Mark, T. Schlick, C. Schütte, R. Skeel (eds.), *New Algorithms for Macromolecular Simulation*. 2006. XVI, 376 pp. Softcover. ISBN 3-540-25542-7

Vol. 50 M. Bücker, G. Corliss, P. Hovland, U. Naumann, B. Norris (eds.), *Automatic Differentiation: Applications, Theory, and Implementations*. 2006. XVIII, 362 pp. Softcover. ISBN 3-540-28403-6

Vol. 51 A.M. Bruaset, A. Tveito (eds.), *Numerical Solution of Partial Differential Equations on Parallel Computers* 2006. XII, 482 pp. Softcover. ISBN 3-540-29076-1

Vol. 52 K.H. Hoffmann, A. Meyer (eds.), *Parallel Algorithms and Cluster Computing*. 2006. X, 374 pp. Softcover. ISBN 3-540-33539-0

Vol. 53 H.-J. Bungartz, M. Schäfer (eds.), *Fluid-Structure Interaction*. 2006. VII, 388 pp. Softcover. ISBN 3-540-34595-7

Vol. 54 J. Behrens, *Adaptive Atmospheric Modeling*. 2006. XX, 314 pp. Softcover. ISBN 3-540-33382-7

For further information on these books please have a look at our mathematics catalogue at the following URL: www.springer.com/series/3527

Monographs in Computational Science and Engineering

Vol. 1 J. Sundnes, G.T. Lines, X. Cai, B.F. Nielsen, K.-A. Mardal, A. Tveito, *Computing the Electrical Activity in the Heart*. 2006. XI, 318 pp. Hardcover. ISBN 3-540-33432-7

For further information on this book, please have a look at our mathematics catalogue at the following URL: www.springer.com/series/7417

Texts in Computational Science and Engineering

Vol. 1   H. P. Langtangen, *Computational Partial Differential Equations*. Numerical Methods and Diffpack Programming. 2nd Edition 2003. XXVI, 855 pp. Hardcover. ISBN 3-540-43416-X

Vol. 2   A. Quarteroni, F. Saleri, *Scientific Computing with MATLAB and Octave*. 2nd Edition 2006. XIV, 318 pp. Hardcover. ISBN 3-540-32612-X

Vol. 3   H. P. Langtangen, *Python Scripting for Computational Science*. 2nd Edition 2006. XXIV, 736 pp. Hardcover. ISBN 3-540-29415-5

For further information on these books please have a look at our mathematics catalogue at the following URL: www.springer.com/series/5151